At Issue

Are Mass Extinctions Inevitable?

Other Books in the At Issue Series:

Alcohol Abuse

Are Textbooks Biased?

Book Banning

Cyberbullying

Distracted Driving

DNA Databases

Do Students Have Too Much Homework?

The Ethics of Capital Punishment

The Ethics of Medical Testing

How Does Advertising Impact Teen Behavior?

Human Waste

Identity Theft

Pandemics

Rebuilding the World Trade Center Site

Should Music Lyrics Be Censored?

Space Exploration

Student Loans

Weapons of War

At Issue

Are Mass Extinctions Inevitable?

Noah Berlatsky, Book Editor

GREENHAVEN PRESS
A part of Gale, Cengage Learning

Detroit • New York • San Francisco • New Haven, Conn • Waterville, Maine • London

Elizabeth Des Chenes, *Managing Editor*

For more information, contact:
Greenhaven Press
27500 Drake Rd.
Farmington Hills, MI 48331-3535
Or you can visit our Internet site at gale.cengage.com

Articles in Greenhaven Press anthologies are often edited for length to meet page requirements. In addition, original titles of these works are changed to clearly present the main thesis and to explicitly indicate the author's opinion. Every effort is made to ensure that Greenhaven Press accurately reflects the original intent of the authors. Every effort has been made to trace the owners of copyrighted material.

LIBRARY OF CONGRESS CATALOGING-IN-PUBLICATION DATA

Are mass extinctions inevitable? / Noah Berlatsky, book editor.
p. cm. -- (At issue)
Includes bibliographical references and index.
ISBN 978-0-7377-5548-0 (hardcover) -- ISBN 978-0-7377-5549-7 (pbk.)
1. Mass extinctions. I. Berlatsky, Noah.
QE721.2.E97A74 2012
576.8'4--dc23

2011036997

Printed in the United States of America
1 2 3 4 5 16 15 14 13 12

FD055

Contents

Introduction

The earth is experiencing a serious increase in extinctions. At the same time, however, scientists have been able to preserve many creatures.

One example is the bald eagle. The bald eagle is well known as America's national bird. In colonial times, bald eagles were extremely numerous in large parts of the United States. However, many ranchers and farmers regarded eagles as pests, and extensive hunting of the birds and destruction of the trees in which the birds nested resulted in a catastrophic drop in their population. By the early 1960s, there were only about thirty-seven hundred bald eagles left, according to David George Gordon's *The Audubon Society Field Guide to the Bald Eagle*. However, the Endangered Species Act of 1973 stepped up protection of the eagles. According to Gordon, "Subsequent federal and state laws have mandated that bald eagle nesting and roosting habitats be managed to maintain and increase eagle numbers to the point of recovery—the threshold where bald eagle populations can sustain themselves." By 1990, Gordon says, "the bald eagle population in the Northwest had increased to 861 nesting pairs—nearly triple the number tallied in 1980." The bald eagle is no longer in danger of extinction, though conservation efforts remain necessary.

More recently, scientists have successfully helped to preserve the Grand Cayman blue iguana, one of the world's largest iguanas. There were fewer than twelve blue iguanas left on Grand Cayman in 2002 when biologists were asked to develop a recovery plan. "The program has met with success," according to a July 19, 2011, article published by ReptileChannel.com. The biologists determined that the iguanas in captivity were not breeding because they were being fed rabbit food and were kept in enclosures that were too small. Their diet was switched to fruits and vegetables, and they were given more room to roam. The biologists also released the iguanas into the wild at an older age, so they would be able to protect themselves against the feral cats that had devastated their population. "With thanks to the captive breeding program, the blue iguana population now numbers around 500 animals and the project says that number should double to around 1,000 in the next two years," ReptileChannel.com concluded.

The bald eagle and the blue iguana are welcome success stories. However, saving individual species sometimes can obscure larger or more intractable problems. For instance, Robert Bloomfield, writing in a January 11, 2010, article in the *Guardian*, notes that preserving biodiversity, or a healthy range of different species of life on earth, "is not about the loss of exotic species which have been the focus of conservation activities by the foundations and trusts of wealthy nations." Instead, Bloomfield argues, the focus needs to be on the biodiversity and resources "that provide the vital needs for the health and wellbeing of us all." In other words, focusing on one or two species to save is not enough; there must be a broad scale approach to preserving natural life.

Some scientists argue that given the vast number of extinctions under way, humans need to think carefully about which species to try to save. Australian researchers have developed a new index to help conservationists better understand how close species are to extinction. These researchers argue that there is a tipping point at which species may no longer be worth preserving—according to them, the tipping point is about five thousand individuals. One of the co-authors of the study, Corey Bradshaw, explains, "we don't have unlimited resources," as quoted in an April 14, 2011, article by John Platt in *Scientific American*. Bradshaw went on to argue that scientists need to prioritize which species to save to achieve the largest ecological impact. "I'd love to save everything," said Bradshaw in the *Scientific American* article. "I just don't think we can," he concluded.

One species scientists have debated about trying to save is the giant panda. Giant pandas are very appealing creatures, and many resources have been poured into protecting them and trying to encourage them to breed in captivity. In a September 23, 2009, article in the *Guardian*, wildlife expert Chris Packam argues that the "millions and millions" spent on pandas could be better used in preserving areas with high biodiversity. "So maybe if we took all the cash we spend on pandas and just bought rainforest with it, we might be doing a better job," he concluded.

However, Mark Wright, of the World Wide Fund for Nature, writing in the same issue of the *Guardian*, contends that focusing conservation efforts on large, appealing animals is a good way to motivate donations and interest from the public in ecological work. In addition, Wright says, preserving the panda habitat helps other creatures, such as "the red panda, golden monkeys, and various birds that are found nowhere else in the world," who live in the same area.

The issue of species preservation is just one of the controversies related to extinction. The viewpoints presented in *At Issue: Are Mass Extinctions Inevitable?* will examine other controversies surrounding species extinction and preservation.

1

Numerous Signs Point to Mass Extinctions

Tracy V. Wilson

Tracy V. Wilson is site director of HowStuffWorks.com.

The earth is experiencing a major wave of extinctions. Extinctions always have occurred, but the current extinction rate is faster than has been the case throughout most of earth's history. Also, the current wave of extinctions appears to be tied to human activity. Human destruction of animal and plant habitat, human-caused global warming, human introduction of invasive species, and other factors appear to be having a major impact on extinction rates.

In January 2009, scientists announced the discovery of a pink iguana on the Galapagos Islands. If you keep your eye on the news, you'll see discoveries like these all the time. As researchers explore the most remote parts of the world, it can seem like new and unusual species crop up every day.

But just as frequent are headlines that warn of species that are dying out, or becoming extinct. The idea that we're in the middle of a widespread, human-caused mass extinction is prevalent in the scientific community.

Researchers have documented a fraction of the species on Earth, so how can we know—or even suspect—that we're in the middle of a mass extinction? Read on to learn about 10 of the biggest warning signs.

Tracy V. Wilson, "Top 10 Signs We Might Be Experiencing a Modern-Day Mass Extinction," *Animal Planet*. http://animal.discovery.com.

A Changing Landscape and Changing Oceans

Dr. Peter Raven of the Missouri Botanical Garden estimates that today's rates of habitat loss will lead to 1,500 extinct species for each million species on Earth.

With a circumference of 24,092 miles (40,075 kilometers) at the equator, Earth seems like a big enough planet for people, animals and plants to share. But according to the American Association for the Advancement of Science, humans are using about half the available land.

Habitat loss is the No. 1 contributor to extinction.

Eleven percent has become farmland, and another 11 percent is used for forestry. Twenty-six percent of the land has become pasture. Another piece, 2 or 3 percent, houses people's homes, businesses, transportation and other services we use to live and work.

All that expansion comes at a price: Habitat loss is the No. 1 contributor to extinction.

While it's easy to picture habitat loss on land, major changes in and near oceans can also play a role in a mass extinction. According to conservationist Tundi Agardy, almost half of the coral reefs on Earth have been destroyed. The same is true of about a third of mangrove forests, which house and protect land and sea animals. Development and industry have led to major changes in the world's coastlines over the last 100 years. Pollution is also a big factor in the survival of marine species.

But habitat loss isn't the only problem—another is overfishing. According to the American Association for the Advancement of Science, two-thirds of the world's marine fisheries are fished close to or beyond their limits. Fish populations can collapse, which can affect the rest of the ecosystem.

On top of that, it's tricky to chart exactly what's going on in the Earth's oceans. Scientists have discovered and classified only a fraction of the species that live there. And sometimes, it's a specific population of fish—not a whole species—that is in danger.

Reduced Biodiversity and Invasive Species

In 1845, potato plants started dying in Ireland. The next six years became known as the Irish Potato Famine. An airborne fungus called *Phytophthora infestans* killed the potatoes, and more than a million people starved. A major factor in the famine was a lack of biodiversity.

Although this might sound like something that couldn't happen in today's industrialized world, a lack of biodiversity can still threaten specific plant and animal species. One of today's examples is the Cavendish banana. The most popular banana choice in the U.S., the Cavendish, like all bananas, is a clone of one parent. And all the Cavendish plants in the world are genetically identical. Two blights currently threaten it—although there will still be other bananas should it become extinct.

The drop in biodiversity also ties in to habitat loss. A smaller space in which to live leads to a smaller population size, which leads to less genetic diversity. Human activities like industrialized farming play a role, too, as large fields of one type of plant replace a diverse population of plants and animals.

Invasive species are contributing to the decline of 42 percent of the world's threatened and endangered species.

If you've ever driven through the American Southeast, you've seen the effects of an invasive species firsthand. Kudzu is a climbing vine that was introduced to the U.S. in the late 1800s. The goal was to control erosion. But kudzu quickly took over parts of the landscape, displacing the plants that used to grow there. Kudzu is just one example. Other invasive species—like pythons in the Florida Everglades—make their way to new habitats unintentionally.

While non-native species can make some contributions to an ecosystem, they also use the food and resources that native species need to live. At the same time, invasive species often have no predators in their new environment. This means their population can grow unchecked. According to paleontologist Niles Eldredge, invasive species are contributing to the decline of 42 percent of the world's threatened and endangered species.

Asteroids and Warming

So far, the Earth has experienced five notable mass extinctions. Scientists don't know the causes for all of them, but asteroid impacts probably made a contribution to at least two. One is the infamous Cretaceous-Tertiary extinction, which led to the end of land-walking dinosaurs. Other factors, including volcanic eruptions, likely played a role in that event as well.

Asteroids collide with the Earth all the time, but most burn up in the atmosphere. The bigger the asteroid, the bigger the threat—but the smaller the chance it will hit the planet. According to NASA [National Aeronautics and Space Administration], an asteroid big enough to seriously affect life on Earth collides with the planet once or twice per million years. NASA and other organizations are surveying the sky to find the locations of all near-Earth asteroids (NEAs) that might pose a threat to the planet.

According to *New Scientist* writer David Chandler, the biggest known threat is an asteroid called 1950 DA. This asteroid has a 1-in-600 chance of colliding with the Earth in 2880—not in our lifetimes, but in the foreseeable future.

A study published in *Nature* suggests that between 15 and 37 percent of the animals researched in the study will either be extinct or on the way to extinction by 2050, depending on the increase in Earth's temperature.

The concept of global warming and whether it's caused by people is still a topic for debate in some circles. But most scientists agree that the temperature is rising—and human activity is the cause.

Critics point out that some animals have adjusted to changes in climate before and even thrived. But other researchers have created different models and equations for exactly how many animals and plants will die out because of global warming. A quarter of all plants and land animals could become extinct, according to a study by Chris Thomas of the University of Leeds. The U.N. [United Nations] Climate Panel states that a 1 degree Celsius (1.8 degree Fahrenheit) increase in temperature will lead to an increasing risk of extinction for 30 percent of species on Earth.

The Fossil Record

Many reports on climate change and extinction focus on the last hundred years or so. But it could be that we're in the middle of an extinction that started about 40,000 years ago, as humans started to settle new areas of the planet. This includes the period a lot of people think of as the last big extinction: the end of the last ice age, when animals like the woolly mammoth and saber-tooth cat became extinct.

It's easy to think of the ice age as its own, isolated event. But people who study the fossil record follow a much longer timeline—one of millions of years instead of thousands. It can take about 100,000 years for a mass extinction to play out in the fossil record. In other words, future scientists may look back on the fossil record and see a pattern that begins tens of thousands of years ago and extends through today. We can see some of this in the fossil record already, and in more recent archaeological evidence.

Extinction is a normal part of life on Earth. And by analyzing the fossil record, scientists have come up with a number known as the background rate of extinction. It's the rate at which species go extinct under normal circumstances, without the influence of asteroid collisions, climate changes, massive volcanic eruptions or other extraordinary factors. According to Peter Douglas Ward, the background rate of extinction is one to five animals per year.

Most researchers agree that today's extinction rate is far beyond the background rate of extinction. According to the U.N. Convention on Biological Diversity, three species are becoming extinct each hour—which is faster than new species are discovered. According to the American Museum of Natural History, this extinction rate is the fastest in the Earth's history.

Biologists Agree

According to the American Association for the Advancement of Science, there were just a few thousand humans on Earth 200,000 years ago. Around 1800, right about the time of the industrial revolution, the Earth's population was 1 billion. The human population reached 6 billion in 1999.

Most researchers agree that today's extinction rate is far beyond the background rate of extinction.

To some minds, the question isn't whether we're in a mass extinction but how big it is. According to the American Museum of Natural History, seven out of 10 biologists surveyed believe the Earth is undergoing a mass extinction.

Researchers on the other side of the fence point out that mass extinctions take time to develop and progress, making it too soon to determine exactly what's happening. Another argument is that a lot of research focuses on negative aspects of invasive species, environmental changes and other factors while ignoring possible benefits.

Over its 4.5-billion-year history, the Earth has been home to a diverse range of plant and animal life. A study of the fossil record uncovers everything from tiny trilobites to enormous dinosaurs. Some fossilized animals and plants bear little resemblance to anything living on Earth today. Others share an uncanny resemblance with their modern descendants.

Regardless of their appearance, the one thing all these fossilized creatures have in common is that they are extinct. Earth has experienced an ongoing cycle of extinction and development throughout its history. Most of these extinctions are separated by 100 million years or so.

The difference between past extinctions and the one most scientists agree is happening today is the cause. Previous extinctions came about because of factors that life on Earth had little control over. But most of the research suggests that an ongoing mass extinction taking place now would have its roots in human activity.

2

Species Extinction Rates Are Overestimated

Stuart Wolpert

Stuart Wolpert is a writer for UCLA Newsroom.

Researchers have determined that models used to calculate rates of extinction are fundamentally flawed. Extinction is occurring at a much slower rate than previously thought. However, researcher Stephen Hubbell emphasizes that extinction is still a serious problem and that ecological preservation is vital. He says better models of rates of extinction are needed to improve preservation efforts.

The most widely used methods for calculating species extinction rates are "fundamentally flawed" and overestimate extinction rates by as much as 160 percent, life scientists report May 19 [2011] in the journal *Nature.*

The Crisis Is Still Serious

However, while the problem of species extinction caused by habitat loss is not as dire as many conservationists and scientists had believed, the global extinction crisis is real, says Stephen Hubbell, a distinguished professor of ecology and evolutionary biology at UCLA [University of California, Los Angeles] and co-author of the *Nature* paper.

"The methods currently in use to estimate extinction rates are erroneous, but we are losing habitat faster than at any time over the last 65 million years," said Hubbell, a tropical forest ecologist and a senior staff scientist at the Smithsonian Tropical Research Institute. "The good news is that we are not in quite as serious trouble right now as people had thought, but that is no reason for complacency. I don't want this research to be misconstrued as saying we don't have anything to worry about when nothing is further from the truth."

Because there are very few ways of directly estimating extinction rates, scientists and conservationists have used an indirect method called a "species-area relationship." This method starts with the number of species found in a given area and then estimates how the number of species grows as the area expands. Using that information, scientists and conservationists have reversed the calculations and attempted to estimate how many fewer species will remain when the amount of land decreases due to habitat loss.

The good news is that we are not in quite as serious trouble right now as people had thought, but that is no reason for complacency.

"There is a forward version when we add species and a backward version when we lose species," Hubbell said. "In the *Nature* paper, we show that this surrogate measure is fundamentally flawed. The species-area curve has been around for more than a century, but you can't just turn it around to calculate how many species should be left when the area is reduced; the area you need to sample to first locate a species is always less than the area you have to sample to eliminate the last member of the species.

"The overestimates can be very substantial. The way people have defined 'extinction debt' (species that face certain extinction) by running the species-area curve backwards is incorrect, but we are not saying an extinction debt does not exist."

A Little More Time

How confident is Hubbell in the findings, which he made with ecologist and lead author Fangliang He, a professor at China's Sun Yat-sen University in Guangzhou and at Canada's University of Alberta?

"100 percent," he said. "The mathematical proof is in our paper."

There were predictions in the early 1980s that as many as half the species on Earth would be lost by 2000.

"Nothing like that has happened," Hubbell said. "However, the next mass extinction may be upon us or just around the corner. There have been five mass extinctions in the history of the Earth, and we could be entering the sixth mass extinction."

Hubbell and He's mathematical proof addresses very large numbers of species and does not answer whether a particular species, such as the polar bear, is at risk of extinction.

"We have bought a little more time with this discovery, but not a lot," Hubbell said.

Hubbell is also the co-founder and co-director of the Center for Tropical Forest Science and is the founder and chairman of the board of the National Council for Science and the Environment (NCSE), an organization with more than 10,000 members and more than 200 universities and professional societies. The NCSE's mission is to improve the science underlying environmental decision-making, and one of the issues the organization addresses is biodiversity conservation and the extinction crisis.

"As we continue to destroy habitat, there comes a point at which we do lose a lot of species—there is no doubt about that," Hubbell said. "However, we have to destroy more habitat before we get to that point."

When a meteor struck the Earth some 65 million years ago, killing the dinosaurs, a fireball incinerated the Earth's forests, and it took about 10 million years for the planet to recover any semblance of continuous forest cover, Hubbell said. The extinctions that humans cause may be as catastrophic, he said, but in different ways.

Humans are already using 40 percent of all the "plant biomass" produced by photosynthesis on the planet, a disturbing statistic because most life on Earth depends on plants, Hubbell noted. Some three-quarters of all species thought to reside on Earth live in rain forests, and they are being cut down at the substantial rate of about half a percent per year, he said.

Hubbell and He used data from the Center for Tropical Forest Science that covered extremely large plots in Asia, Africa, South America and Central America in which every tree is tagged, mapped and identified—some 4.5 million trees and 8,500 tree species. Many of these tree species are very rare. If they go extinct, so will the animals that depend on them.

We also need much deeper thought about how we can estimate the extinction rate properly to improve the science behind conservation planning.

"We need much better data on the distribution of life on Earth," he said. "We need to rapidly increase our understanding of where species are on the planet. We need citizens to record their local biodiversity; there are not enough scientists to gather the information. We also need much deeper thought about how we can estimate the extinction rate properly to improve the science behind conservation planning. If you don't know what you have, it is hard to conserve it."

Hubbell and He have worked together for more than 25 years through the Center for Tropical Forest Science. The research was federally funded by the National Science Foundation, NASA [National Aeronautics and Space Administration], and the Natural Sciences and Engineering Research Council of Canada. While the current research estimates that extinction rates have been overreported by as much as 160 percent, Hubbell and He plan in future research to investigate more precisely how large the overestimates have been.

If we don't take steps to preserve animals and plants that we care about, they are going to be gone.

The Loss of a Conservation Ethic

Hubbell encourages the public to spend more time enjoying nature, "especially if it's going to be here today, gone tomorrow. If we don't take steps to preserve animals and plants that we care about, they are going to be gone.

"When I was a kid," he said, "I spent a lot of time doing non-macho things like collecting butterflies and turning over rocks. The only way we're going to save nature is by making sure future generations experience nature. People who have never seen wild nature don't miss it and don't realize how impoverished their lives have become due to its loss. I worry about the loss of a conservation ethic among the public. Go to the tropics. Experience a rain forest—while you still can."

3

Study: 6th Mass Extinction Looms But Preventable, Study Says

Stephanie Pappas

Stephanie Pappas is a senior writer at MSNBC.com.

Earth faces a major extinction event on a scale experienced only five times before in the history of the planet. The result of extinctions could be catastrophic and unpredictable. However, because extinctions are being driven by human actions—such as climate change and habitat destruction—it should be possible for humans to prevent them. The issue is one of political will. While some scientists see no hope that governments will address these problems, others are optimistic that policy changes can prevent catastrophe once the scale of the problem becomes apparent.

Are humans causing a mass extinction on the magnitude of the one that killed the dinosaurs?

The answer is yes, according to a new analysis—but we still have some time to stop it.

Mass extinctions include events in which 75 percent of the species on Earth disappear within a geologically short time period, usually on the order of a few hundred thousand to a couple million years. It's happened only five times before in the past 540 million years of multicellular life on Earth. (The last great extinction occurred 65 million years ago, when the dinosaurs were wiped out.) At current rates of extinction, the study found, Earth will enter its sixth mass extinction within the next 300 to 2,000 years.

"It's bittersweet, because we're showing that we have this crisis," study co-author Elizabeth Ferrer, a graduate student in biology at the University of California, Berkeley, told LiveScience. "But we still have time to fix this."

Others aren't so optimistic that humans will actually do anything to stop the looming disaster, saying that politics is successfully working against saving species and the planet.

It's bittersweet, because we're showing that we have this crisis. . . . But we still have time to fix this.

The 6th Extinction

Species go extinct all the time, said Anthony Barnosky, the curator of the Museum of Paleontology at UC Berkeley and another co-author of the paper, which appears in this week's issue of the journal *Nature*. But new species also evolve constantly, meaning that biodiversity usually stays constant. Mass extinctions happen when that balance goes out of whack. Suddenly, extinctions far outpace the genesis of new species, and the old rules for species survival go out the window. [Read: Mass Extinction Threat: Earth on Verge of Huge Reset Button?]

"If the fossil record tells us one thing, it's that when we kick over into a mass extinction regime, results are extreme, they're irreversible and they're unpredictable," David Jablonski, a paleontologist at the University of Chicago who was not involved in the study, told LiveScience. "Factors that promote success and survival during normal times seem to melt away."

Everyone knows that we now lose many species a year, Barnosky said. "The question is, 'Is the pace of extinction we're seeing today over these short time intervals usual or unusual?'"

Answering the question requires stitching together two types of data: that from the fossil record and that collected by conservation biologists in the modern era. They don't always match up well. For example, Barnosky said, fossils tell us lots about the history of clams, snails and other invertebrates. But in the modern world, biologists have only assessed the extinction risk for 3 percent of known species of such invertebrates. That makes comparisons tough.

The fossil record also presents a blurrier history than today's yearly records of species counts. Sparse examples of a species may be distributed across millions of years of fossil history, the researchers wrote, while modern surveys provide dense samples over short periods of time. And even the best source of modern data—the International Union for the Conservation of Nature Red List of threatened and endangered species—has catalogued the conservation status of less than 2.7 percent of the 1.9 million named species out there.

Even the best source of modern data . . . has catalogued the conservation status of less than 2.7 percent of the 1.9 million named species out there.

Coming Crisis

The researchers worked to combine these two sources of data, Ferrer said, taking a conservative approach to filling in gaps and estimating future directions. They found that the overall rate of extinction is, in fact, between three to 80 times higher than non-mass extinction rates. Most likely, species are going extinct three to 12 times faster than would be expected if there were no crisis, Ferrer said.

That gives Earth between three and 22 centuries to reach the point of mass extinction if nothing is done to stop the problem. (The wide range is a factor of the uncertainty in the data and different rates of extinction found in various species.)

The good news, Barnosky said, is that the total loss so far is not devastating. In the last 200 years, the researchers found, only 1 to 2 percent of all species have gone extinct.

The strongest evidence for comparison between modern and ancient times comes from vertebrate animals, Barnosky said, which means there is still work to do collecting better data for more robust comparisons with better invertebrate data. But, he said, the research "shows absolutely without a doubt that we do have this major problem."

Back from the Brink?

The culprits for the biodiversity loss include climate change, habitat loss, pollution and overfishing, the researchers wrote.

"Most of the mechanisms that are occurring today, most of them are caused by us," Ferrer said.

So can we fix it? Yes, there's time to cut dependence on fossil fuels, alleviate climate change and commit to conservation of habitat, the study scientists say. The more pressing question is, will we?

The culprits for the biodiversity loss include climate change, habitat loss, pollution and overfishing, the researchers wrote.

Barnosky and Ferrer both say they're optimistic that people will pull together to solve the problem once they understand the magnitude of the looming disaster. Jablonski puts himself into the "guardedly optimistic category."

"I think a lot of the problems probably have a lot more to do with politics than with science," Jablonski said.

That's where Paul Ehrlich, the president of the Center for Conservation Biology at Stanford University and author of "The Population Bomb" (Sierra Club-Ballantine, 1968), sees little hope.

"Everything we're doing in Washington [DC] today is working in the wrong direction," Ehrlich, who was not involved in the research, told LiveScience. "There isn't a single powerful person in the world who is really talking about what the situation is . . . It's hard to be cheery when you don't see the slightest sign of any real attention being paid."

Other researchers take an upbeat view.

"If we have a business-as-usual scenario, it is pretty grim, but it isn't yet written," Stuart Pimm, a professor of conservation ecology at Duke University who was not involved in the research, told LiveScience in a phone interview from Chile, where he was doing fieldwork.

In 2010, Pimm said, the United Nations declared the International Year of Biodiversity. According to a UN statement, the 193 countries involved agreed to protect 17 percent of Earth's terrestrial ecosystems and 10 percent of marine and coastal areas. Some types of ecosystems still lag behind. Pimm said, but there is reason for hope.

"I hope that this will alert people to the fact that we are living in geologically unprecedented times," Pimm said. "Only five times in Earth's history has life been as threatened as it is now."

4

Mass Plant Extinctions Show the Threats from Human Exploitation

Juliette Jowit

Juliette Jowit is the environmental editor at the Observer, *Britain's oldest Sunday newspaper.*

A study has found that 20 percent of plant species on earth are at risk of extinction. Although this number is lower than previous estimates, many experts remain seriously concerned about the threat of another mass extinction. Plants are vital to the biodiversity of animal species, and they provide humans with food, water, and medicine. Plants also help to mitigate climate change. The threat to plant species, therefore, is a threat to all life on earth. Scientists argue that we must move away from exploiting plants and should instead begin to care for plants more carefully.

One in five of the world's plant species—the basis of all life on earth—are at risk of extinction, according to a landmark study published today [September 29, 2010].

One-Fifth of Plants Under Threat

At first glance, the 20% figure looks far better than the previous official estimate of almost three-quarters, but the announcement is being greeted with deep concern.

The previous estimate that 70% of plants were either critically endangered, endangered or vulnerable was based on what scientists universally acknowledged were studies heavily biased towards species already thought to be under threat.

Today the first ever comprehensive assessment of plants, from giant tropical rainforests to the rarest of delicate orchids, concludes the real figure is at least 22%. It could well be higher because hundreds of species being discovered by scientists each year are likely to be in the "at risk" category.

Plants are the basis of life.

"We think this is a conservative estimate," said Eimear Nic Lughadha, one of the scientists at Kew Gardens in west London [United Kingdom] responsible for the project.

The plant study is also considered critical to understanding the level of threat to all the natural world's biodiversity, said Craig Hilton-Taylor of the International Union for the Conservation of Nature (IUCN), which runs the world's offical "red list" of threatened species. "Plants are the basis of life, and unless we know what's happening to plants it has many implications," said Hilton-Taylor.

The results will be presented to world leaders meeting at Nagoya in Japan in October [2010] to discuss the world's biodiversity crisis, along with new red lists for vertebrates and several groups of the planet's millions of invertebrate species.

Nurturing Plants

"This is a base point," said Nic Lughadha. "What we do from now is going to lead to the future of plants. We need to challenge the idea that plants are there to be exploited by us, we need to move to a system where we're nurturing plants much more carefully [and] actively taking steps to conserve them."

Politicians and conservation experts will also be told that by far the biggest threat to plants is human—rather than natural—causes, especially intensive agriculture, livestock grazing, logging and infrastructure development.

Caroline Spelman, the environment secretary [of the United Kingdom], who will travel to Japan for the final talks, said the results were deeply troubling. She added: "Plant life is vital to our very existence, providing us with food, water, medicines, and the ability to mitigate and adapt to climate change."

Scientists randomly selected 7,000 species from across the major plant groups as a representative sample of the estimated 380,000–400,000 so far known to science. Of these, 3,000 were found to have too little information to begin making a proper assessment—a result that was expected and so built into the selection process.

We need to move to a system where we're nurturing plants much more carefully.

The remaining 4,000 species were assessed and the level of risk based on a combination of the absolute number of plants estimated in the wild, the known decline, and the total area in which they are thought to live.

Of the 4,000, 63% were found to be of "least concern", 10% near threatened, 11% vulnerable, 7% endangered and 4% critically endangered. Another 5% were rated "data deficient".

The proportion of plant species deemed at-risk is similar to that of the IUCN's red list for mammals, worse than that for birds (less than 10% at-risk) and better than the number for amphibians (more than a quarter under threat).

Nearly two-thirds of threatened plant species are found in tropical rainforests, five times the proportion for the nearest other habitats—rocky areas, temperate forests and tropical dry forests. This is because of their huge density of biodiversity and the widespread risks of logging and clearance for other agriculture, said analysts.

Previously the red list for plants contained assessments for a greater number of plants—about 12,873 or 3% of known species—but was not considered representative because scientists had focused on at-risk species so that they could get attention and funding for conservation.

Nearly two-thirds of threatened plant species are found in tropical rainforests.

The assessment was done using experts and collections at the herbaria at Kew Gardens, the Natural History Museum in London and Missouri Botanical Garden in the US, plus specialist experts from the IUCN.

Six of the Endangered Plants

Wollemi pine (*Wollemia nobilis*)—critically endangered

The wollemi pine was discovered in 1994 in Wollemi National Park, Australia, and fewer than 50 mature individuals are known. Its long-term regeneration from seed is unknown but seems doubtful due to competition with other trees. Its small size and limited range means it is at risk from any chance event such as fire or the spread of disease.

Common snowdrop (*Galanthus nivalis*)—near threatened

The common snowdrop was once widely distributed in the east Carpathian mountains in central and eastern Europe. Although it is widely naturalised, including in the UK [United Kingdom], during the past decade its native distribution has been considerably reduced, due mainly to habitat loss through the increase in residential developments and recreational land use.

Rosewood (*Dalbergia andapensis*)—critically endangered

D. andapensis is a species of rosewood, a highly valued timber used in the production of fine furniture and musical instruments. It is estimated that 52,000 tonnes of rosewood and ebony were logged in north-east Madagascar in 2009, and this habitat is itself under threat from conversion to agriculture for a growing rural population.

Wood bitter-vetch (*Vicia orobus*)—least concern

Wood bitter-vetch is a rare species found through much of western Europe, including the British Isles, at woodland margins, field edges and rocky places, often on limestone. In Ireland it is considered to be threatened as a result of habitat loss, and is being protected by the National Botanic Gardens of Ireland.

Whited's milkvetch (*Astragalus sinuatus*)—critically endangered

Whited's milkvetch is restricted to a tiny area of the state of Washington, USA. Its dry hillside habitat is threatened by invasive, non-native species, by grazing and by agriculture. Seeds have been collected and banked by the Berry Botanic Garden Seed Bank for Rare and Endangered Plants of the Pacific Northwest and the Miller Seed Vault, University of Washington Botanic Gardens.

Encephalartos altensteinii—vulnerable

E. altensteinii is found in coastal regions of the eastern cape, South Africa, where the number of individuals has declined by more than 30% in the past 50 years. Large numbers have been removed from its native habitat, including 438 plants in one poaching incident in 1995, mainly by horticultural collectors for pot plants or medicinal use.

5

Human Consumption of Plants and Animals Can Lead to Conservation

Kim Severson

Kim Severson is a food writer for the New York Times *and the author of* Spoon Fed: How Eight Cooks Saved My Life.

Several groups are working to preserve animal and plant species that have been part of regional American diets. By encouraging chefs and farmers to raise the species as foods, the groups have helped to bring back varieties of endangered potatoes, chickens, and cattle. This strategy works well with many edible species; however, there are some species whose numbers have been so reduced that harvesting them for food might push them toward extinction.

Some people would just as soon ignore the culinary potential of the Carolina flying squirrel or the Waldoboro green neck rutabaga. To them, the creamy Hutterite soup bean is too obscure and the Tennessee fainting goat, which keels over when startled, sounds more like a sideshow act than the centerpiece of a barbecue.

Eating Animals to Save Them

But not Gary Paul Nabhan. He has spent most of the past four years compiling a list of endangered plants and animals that were once fairly commonplace in American kitchens but are now threatened, endangered or essentially extinct in the marketplace. He has set out to save them, which often involves urging people to eat them.

Mr. Nabhan's list, 1,080 items and growing, forms the basis of his new book, an engaging journey through the nooks and crannies of American culinary history titled *Renewing America's Food Traditions: Saving and Savoring the Continent's Most Endangered Foods.*

Mr. Nabhan's book is part of a larger effort to bring foods back from the brink by engaging nursery owners, farmers, breeders and chefs to grow and use them.

The book tells the stories of 93 ingredients both obscure (Ny'pa, a type of salt grass) and beloved (the Black Sphinx date), along with recipes that range from the accessible (Centennial pecan pie) to the challenging (whole pit-roasted Plains pronghorn antelope).

To make the list, an animal or plant—whether American eels, pre-Civil War peanuts or Seneca hominy flint corn—has to be more than simply edible. It must meet a set of criteria that define it as a part of American culture, too. Mr. Nabhan's book is part of a larger effort to bring foods back from the brink by engaging nursery owners, farmers, breeders and chefs to grow and use them.

"This is not just about the genetics of the seeds and breeds," said Mr. Nabhan, an ethnobotanist and an expert on Native American foods who raises Navajo churro sheep and heritage crops in Arizona. "If we save a vegetable but we don't save the recipes and the farmers don't benefit because no one eats it, then we haven't done our work."

The Ark of Taste

He organized his list into 13 culinary regions that he calls nations, borrowing from Native American and other groups. The

Pacific Coast from California to northern Mexico is acorn nation. Its counterpart on the mid-Atlantic coast is crab cake nation. Moose nation covers most of Canada. New Yorkers, for the record, live in clambake nation.

His work is based on extensive trips around the country, where he listened to old-timers and cataloged hundreds of hard-to-find plants and animals, like the finicky Datil chili pepper (originally from Cuba), the Bronx grape and the long-stemmed Harrison cider apple from New Jersey.

"The daunting thing is that so much about American traditional foods comes out of people's heads and isn't in any book," he said. He had little trouble getting people to share their knowledge. "This to them is like a baseball fan talking about the Yankees. They just know all the details."

Mr. Nabhan engaged seven culinary, environmental and conservation groups to help him identify items for the list and return them to culinary rotation.

He acted like a broker for the groups, some of which had been trying to save traditional food for decades. Organizations including the Seed Savers Exchange and the American Livestock Breeds Conservancy contributed suggestions for the list. Then, leveraging the rising interest in regional food, he engaged hundreds of chefs, farmers and curious eaters to grow and cook some of the lost breeds and varieties.

Farmers are often more concerned with innovating and crossbreeding than in preserving cultural traditions or encouraging biological diversity.

Leading the way are members of the gastronomic group Slow Food U.S.A., which assesses whether foods on Mr. Nabhan's list are delicious and meaningful enough in the communities where they originated to be worth reviving and promoting. Foods that do become part of what the group calls its Ark of Taste.

The Chefs Collaborative, a group of more than 1,000 professional cooks and others dedicated to sustainable cuisine, willingly signed on, too. Several members incorporated traditional ingredients into modern restaurant dishes, holding a series of picnics last year to show off their work.

And everyone in Mr. Nabhan's alliance tried to encourage farmers and ranchers to grow the seeds and the breeds, promising to deliver buyers if they did.

That is the most complicated part of reviving traditional food, said Makalé Faber Cullen, a cultural anthropologist with Slow Food U.S.A. who contributed to the book. Farmers are often more concerned with innovating and crossbreeding than in preserving cultural traditions or encouraging biological diversity.

"That's where the tension lies in this project," she said. "A lot of times products fall into disuse because farmers themselves decide they are not worthy of the marketplace. A farmer will say, I don't want to grow out that tomato anymore. I want something with thicker skin."

Some of the items on the list, like Ojai pixie tangerines and Sonoma County Gravenstein apples, were well on their way back before Mr. Nabhan came along. But other foods are enjoying a renaissance largely as a result of the coalition's work.

The Makah ozette potato, a nutty fingerling with such a rich, creamy texture that it needs only a whisper of oil, is one of the success stories. It is named after the Makah Indians, who live at the northwest tip of Washington state and have been growing the potatoes for more than 200 years.

The Seattle chapters of Slow Food and the Chefs Collaborative adopted the rare potato. In 2006, Slow Food passed out seed potatoes to a handful of local farmers and gardeners, and chefs like Seth Caswell at the Stumbling Goat Bistro in Seattle began putting them on the menu.

Mr. Caswell says they are delicious roasted with a little hazelnut oil for salads or cut into wedges to go with burgers made with wagyu beef and Washington State black truffle oil.

Successful Revivals

There have been other revivals, the moon and stars watermelon and the tepary bean among them. The effort to reintroduce heritage turkeys to the American table was a precursor to the work of Mr. Nabhan and his collaborators.

The meaty Buckeye chicken, with its long legs suitable for ranging around, is considered one of five most endangered chicken breeds. Last year over 1,000 chicks were hatched and delivered to breeders, Mr. Nabhan said.

The idea of eater-based conservation . . . works well for agricultural products and some wild foods like clams that benefit from regular harvesting.

Justin Pitts, whose family has raised Pineywoods cattle in southern Mississippi for generations, credits the coalition with saving those animals. The small, lean cattle that provide milk, meat and labor spent centuries adapting to the pine barrens of the deep south, raised by families who can trace their herds back as far as anyone can remember. There are less than a dozen of those families left, and at one point the number of pure Pineywoods breeding animals fell to under 200. In the past few years, it has grown to nearly 1,000.

Mr. Pitts, who has "90 head if I can find them all," sells New York strips and other cuts at the New Orleans farmers' market and to chefs.

"I can't raise cattle fast as they eat them," he said.

He supports the notion that you've got to eat something to save it.

"If you're keeping them for a museum piece," he said, "you've just signed their death warrant."

But Mr. Nabhan doesn't want people to eat everything on his list. The idea of eater-based conservation, which holds that to save something, one has to eat it, works well for agricultural products and some wild foods like clams that benefit from regular harvesting. For some wild species, however, like the foot-long, pink-fleshed Carolina flying squirrel, a harvest would create too much pressure on a tiny population.

The squirrels used to make regular appearances in Appalachian game-meat stews. But as their forests declined, so did the squirrel population; they are now on state and federal endangered species lists. Even if catching them were legal, Mr. Nabhan says a trapper would be hard-pressed to bag more than half a dozen a season.

Because the squirrel was once so important to the diets of North Carolina and east Tennessee, Mr. Nabhan included it on his list, along with a recipe for the thick vegetable stew called Kentucky burgoo.

It calls for corn, lima beans, spring water and two pounds of cubed and fried squirrel meat. Just don't use flying squirrel. At least not yet.

6

Humans Have Created Conditions That Cause Amphibian Extinctions

NZFROG

NZFROG is a website organized by New Zealand's two main centers of frog research and conservation—the University of Otago Frog Research Group and the Victoria University of Wellington Frog Research Group.

Amphibians worldwide have been facing extreme declines. Currently, a third of amphibians are threatened with extinction. This is a serious problem, because frogs are an important part of the food web, consuming insects and serving as vital prey themselves. Frogs are threatened by global warming and many human activities, including habitat destruction, pollution, the introduction of predators, and the sale of frogs as food and pets.

After many years of worrying about amphibian declines and trying to pinpoint the exact cause of the problem, scientists are now faced with an even more serious problem. The declines have become so severe that scientists are now watching their study animals become extinct. We have now moved into the phase of amphibian extinctions rather than studying amphibian declines and 32% of all amphibians are threatened with extinction. When a whole group of a particular type of animal starts to disappear then we need to worry. Amphibians play an essential part in the food web of life as top insectivores and prey to many other animals, and if you remove this important link then one can only guess about the ramifications, but there is no doubt that they will be serious.

The text below contains information on the main causes of amphibian extinctions. . . .

What Is the Cause?

The issue of declining frog populations was first highlighted in 1989 when a rather large group of over 1400 [researchers] from 60 countries throughout the world, descended upon the delightful town of Canterbury in the UK [United Kingdom] for the First World Congress of Herpetology. It was at this meeting that a disturbing number of researchers reported apparent declines in their study populations and of growing concern were the reports of declining populations from seemingly pristine habitats. We are now nearly 20 years down the line, with over 250 publications being produced relating to global amphibian declines and as yet we are not much further ahead in identifying the cause. Throughout the world over 200 amphibian species have experienced recent population declines, with reports of at least 32 species extinctions.

So what could be the causes of declining amphibian populations? . . .

Habitat Destruction

Far too often we see depressing sights of blatant habitat destruction, and with the ever increasing human population, habitat destruction and fragmentation will become even more commonplace. It is now estimated that humans have altered between one third and one half of the Earth's land surface.

Forestry and agriculture take a major toll on frog populations. In fact habitat destruction and alteration is the main objective of most farmers. It has also been demonstrated that roads and agricultural fields are significant barriers to terrestrial arthropods [such as insects and arachnids] and small mammals and the same is probably true for amphibians. Mining also constitutes a major threat to amphibians, not only through the associated water pollution but also through the mechanical destruction of frogs themselves as well as their habitat.

Frogs vary a great deal in their tolerance to acid water. Tadpoles tolerate acidity better than embryos (fertilisation being the most sensitive developmental stage to acidity), and tolerance increases with age. Many frogs breed in temporary pools that fill up with rain. The acidity of these pools may be strongly influenced by the acidity of the rain and therefore may be considerably more acidic than that of nearby ponds and lakes. Acidification will probably constitute a major threat to our frogs.

Forestry and agriculture take a major toll on frog populations.

Aluminium, cadmium, copper, zinc and iron are all toxic to amphibians. We can also infer from studies on fish that nickel, lead, and manganese will have damaging effects on frog populations. It has also been demonstrated that large amounts of lead, from car exhaust gases, are deposited into major water bodies. We can expect heavy metals to play a leading role in the downfall of our frogs.

Organic herbicides and pesticides often cause developmental abnormalities or fatalities. A report in 1997 demonstrated that the widely used and apparently safe herbicide "Roundup" was extremely toxic to tadpoles and adult frogs. This herbicide (in its new and apparently safe form) is still widely used by farmers, foresters and gardeners of New Zealand. Obviously, the insecticides DDT [dichlorodiphenyltrichloroethane] and Dieldrin are dangerously toxic at very low levels and at present we know very little about rates of degradation under field conditions. Certain chemicals such as DDT and PCBs [polychlorinated biphenyls] mimic hormones and can cause sterility in a number of frog species and may possibly influence the fertility of humans.

Atrazine, a major herbicide used in New Zealand has been blamed for the chemical castration of frogs (and other animals, possibly humans). . . .

Global Warming

The greenhouse effect was described more than 100 years ago and its effect on our planet was well understood. Global climate change can affect our frog populations directly or indirectly.

Direct effects—The timing of breeding by amphibians is governed by environmental factors such as temperature. It is thought that if global warming occurs then the breeding season of frogs will change and frogs will start breeding earlier. This is exactly what is found in many populations, with the frogs coming out of hibernation earlier being more susceptible to sudden cold changes in the weather.

Current general circulation models predict that a doubling of the present carbon dioxide will occur in the second half of this century predicting an increase of the mean Earth's surface temperature by 4 C [degrees Celsius]. This will result in an increase in sea level by over 2 m [meters] which apart from destroying many areas of human habitation will inundate most of the world's coastal wetlands. From our amphibian point of view it is all doom and gloom—there will be an alteration of rainfall patterns, more frequent and intense droughts, and this coupled with acid rain and pollutants only exacerbates the problems faced by our moist friends.

Indirect effects—It is possible that these adverse climatic conditions may have very subtle effects on frogs, such as depression of immune function causing the frogs to become more susceptible to disease. In addition there may be even more complex subtle effects where the lower pond water levels will cause the embryos to be exposed to more UV [ultraviolet] light from the sun causing them to be even more susceptible to fungal attacks.

It has been suggested that the widespread distribution of amphibian declines involves global agents such as an increase in ultra-violet radiation (in particular UV-B at a wavelength of 300 nm [nanometers]. UVB radiation is no doubt very harmful to many species of amphibians, however the literature surrounding the effects of UV on frog populations is highly controversial and requires much more work before this can be determined to be a serious threat.

Predators, Diseases, and Food

Predators—In the United States, Australia and New Zealand there have been several populations of amphibians that have declined as a direct result of predation by freshwater crayfish, bullfrogs, cane toads, and rats. In this country [New Zealand] the fish commonly known as mosquito fish have been found to be major predators on tadpoles and native fish and I suspect that this invasive species will play a major role in the demise of local populations of introduced frogs. In Australia there is a movement to change the name of this fish to "plague minnow" as they have been shown to be a voracious predator and extremely hardy of salty, acidic or hot conditions—I even found them swimming around in some thermal pools at Rotorua [New Zealand]. It is thought that the major agent of decline of our native frogs was the introduction of mammalian predators such as kiore, mice and the Mustelids (stoats, weasels, etc).

Diseases—Diseases have been implicated in many amphibian declines. Viral, bacterial and fungal diseases have been reported as the major agents in the declines of many species. In New Zealand a recent outbreak of the chytrid (pronounced 'kit-rid') fungus caused a mass die-off of Archey's frog in the Coromandel, resulting in an emergency meeting of the Native

Frog Recovery group. Archey's frog is now listed as Nationally Critical and steps are in place to prevent its extinction.

Another possible cause of global declines is the collection of frogs as a food source. Africa, United States, Germany, France and the Netherlands consume millions of frogs annually. For example, the United States has been recorded as importing more than 3 million kg [kilograms] of frog meat per year. That is the equivalent of approximately 26 million frogs!

The main exporters of frogs are India, Indonesia and Bangladesh. When these frogs are captured on a large scale their usual prey items multiply out of control. Consequently, this results in the extensive use of insecticides, at a considerable cost, often at a higher cost than the revenue the farmers receive from exporting frogs legs. By using vast quantities of insecticides they cause considerable damage to the delicate ecosystem, including destruction of further frog populations. Presently the consumption of frogs in this country is only on a very small scale, being mainly confined to visiting French scientists!

The United States has been recorded as importing more than 3 million kg of frog meat per year. That is the equivalent of approximately 26 million frogs!

The pet trade, which is a multi-million dollar business, also plays an important role and generally the rarer the species the higher the demand and the rarer the frog becomes.

Frogs caught in the wild for the pet trade are often kept in crowded cages awaiting final shipment to the United States. These animals are kept in very stressful conditions, and the ones that do make it alive to the United States usually only last a couple of months in captivity before being unceremoniously flushed down the toilet.

New Zealand has one of the rarest frogs in the world and the demand for this species in the pet trade or by fanatical collectors is very high, and it is only due to the hard work of the Department of Conservation and the frogs inaccessible habitat that enable this frog to survive today.

The whole picture itself of declining populations is not an easy one to unravel as many of these causative factors interact and intensify the problems facing the frogs. For example, while only large doses of UVB are necessary to cause substantial mortality, smaller doses may increase the amount of stress on tadpoles making them more susceptible to diseases. Also, acidity has been shown to seriously affect the toxicity of different metals on different species of frogs. Consequently, all these factors are interrelated and a complex web of stresses may be causing the demise of many populations, and this intricate nature prevents the isolation of one causative factor.

A complex web of stresses may be causing the demise of many populations.

Frogs Are Important

Why should we be concerned about frogs, they are lowly creatures after all? As they are major predators of insects they fulfil an important role in the food chains and without them insects can multiply out of control causing considerable damage to crops and a dramatic increase in insect borne diseases such as encephalitis and malaria. As we have seen most frogs have a biphasic life cycle, where eggs laid in water, develop into tadpoles and these live in the water until they metamorphose into tiny replicas of the adults. This fact, coupled with being covered by a semi-permeable skin, makes frogs particularly vulnerable to pollutants and other environmental stresses. Consequently frogs can be used as environmental sentinels or biomonitors and act as an early warning system for the quality of the environment and the potential threats to other animals including ourselves. In addition frogs throughout the world provide a valuable source of food for humans and other animals.

7

A Fungal Disease Is One of the Main Threats Causing Amphibian Extinctions

Amphibian Ark

Amphibian Ark, an organization that strives to preserve amphibians, is made up of three principal partners: the World Association of Zoos and Aquariums (WAZA), the International Union for Conservation of Nature/Species Survival Commission (IUCN/SSC), Conservation Breeding Specialist Group (CBSG), and the IUCN/SSC Amphibian Specialist Group (ASG).

The bacteria Bd *causes the disease chytridiomycosis in many amphibians. This disease is the most devastating infectious disease ever encountered among vertebrate species. It causes massive die-offs of amphibians, and can quickly push a given species toward extinction. The exact origin of this disease is unknown, although it has been spread by humans trading in amphibians. Habitat loss caused by human action continues to be the largest threat to most amphibian species, but* Bd *can cause rapid extinction and is extremely difficult to combat once it is established in a population.*

A "chytrid" is a type of fungus (Phylum Chytridiomycota) and there are approximately 1,000 different chytrid species that live exclusively in water or moist environments. The chytrids are among the oldest (most primitive) types of fungi and until very recently were considered members of the Kingdom Protista (and therefore thought to be more closely related to single celled organisms like protozoa). Most chytrids are saprobes meaning that they feed on dead and rotting organic matter. Other chytrids are parasites that live on plants or invertebrate animals. In 1999, a new species of chytrid was described that infects the skin of amphibians and was named *Batrachochytrium dendrobatidis* or "*Bd*" for short. Although the name *Batrachochytrium* is hard for even many scientists to pronounce, it roughly translates to mean "frog chytrid". *Bd* is unusual because it is the only chytrid that is a parasite of a vertebrate animal (amphibians specifically; *Bd* has not been observed to infect other vertebrates such as reptiles, birds or mammals).

Why Is *Bd* Important?

Bd is a very important chytrid fungus because it appears to be capable of infecting most of the world's approximately 6,000 amphibian species and many of those species develop the disease chytridiomycosis which is linked to devastating population declines and species extinctions. In fact, infection with *Bd* has been called "the worst infectious disease ever recorded among vertebrates in terms of the number of species impacted, and its propensity to drive them to extinction. Amphibian population declines due to chytridiomycosis can occur very rapidly—sometimes over a just a few weeks and disproportionately eliminate species that are rare, specialized and endemic (e.g. those species that are most unique). Because of these characteristics—rapid progression of population declines and loss of very important amphibian species—urgent mobilization of efforts to preserve amphibian species are required.

Chytridiomycosis ("Mycosis" = disease caused by a fungus) is the disease that occurs when an amphibian is infected with large numbers of the *Bd* fungus. Infection with *Bd* occurs inside the cells of the outer skin layers that contain large amounts of a protein called "keratin". Keratin is the material that makes the outside of the skin tough and resistant to injury and is also what hair, feathers and claws are made of. With chytridiomycosis, the skin becomes very thick due to a microscopic change in the skin that pathologists call "hyperplasia and hyperkeratosis". These changes in the skin are deadly to amphibians because—unlike most other animals—amphibians "drink" water and absorb important salts (electrolytes) like sodium and potassium through the skin and not through the mouth. Abnormal electrolyte levels as the result of *Bd*-damaged skin cause the heart to stop beating and the death of the animal. Other amphibians like the lungless salamanders, use the skin to breathe and skin changes due to chytridiomycosis could interfere with this function causing suffocation.

Infection with Bd *[by amphibians] has been called "the worst infectious disease ever recorded among vertebrates in terms of the number of species impacted, and it's propensity to drive them to extinction."*

Investigating Resistance

Not all amphibian species that are infected with *Bd* become sick or die. These species like the American bullfrog and the African clawed frog are said to be "resistant" to chytridiomycosis. Resistant species are a major concern because they are carriers of *Bd* (like a "Typhoid Mary" [infection carrier/spreader]) that can move the fungus to new locations and expose new populations of amphibians that are "susceptible" or more likely to become sick with lethal chytridiomycosis.

The reason why some amphibian species are resistant to chytridiomycosis is an area of very active scientific research. If we can understand why some species are resistant, it might be possible to develop methods to control chytridiomycosis in amphibian populations that experience devastating population declines. Some of the mechanisms that could explain species resistance to chytridiomycosis are:

- The presence on the skin of specific types of symbiotic bacteria that discourage the growth of *Bd*. Amphibians or amphibian populations that normally have large numbers of these bacteria in the skin might be more resistant to developing chytridiomycosis.
- The production by the poison glands in amphibian skin of chemicals called "antimicrobial peptides" that discourage the growth of *Bd*. Specific types, combinations or amounts of antimicrobial peptides might help some species to be more resistant to chytridiomycosis.
- Some amphibian species or populations may have genetic resistance to the development of chytridiomycosis by mechanisms that are not yet understood.

If we can understand why some species are resistant, it might be possible to develop methods to control chytridiomycosis.

Other scientists study why some populations of amphibians succumb to chytridiomycosis while other populations of the same species persist. In addition to things like the presence of symbiotic bacteria or differences in skin peptide composition, some potential explanations include:

- Environmental differences between populations such as temperature, humidity or water flow patterns. For instance, some of the most important amphibian population declines associated with chytridiomycosis have occurred at high elevation locations that have a cool temperature range (< 25°C or 77°F) that is most optimal for the growth of *Bd*.

- Differences in virulence between different types or "strains" of the *Bd* fungus. The term virulence refers to the ability of the fungus to cause disease in amphibians. A type of *Bd* that is "highly virulent" easily makes amphibians sick, but another type of *Bd* that has "low virulence" makes fewer animals sick or results in less severe disease.

There is not a single explanation for why an amphibian population succumbs or does not succumb to chytridiomycosis and in most cases multiple factors are probably at work to result in a particular outcome.

Some amphibian populations experience devastating mass mortality events due to chytridiomycosis where most of the population succumbs to the disease, but a small number of animals remain or "persist" in the population. At this time it is unknown if these "persistent populations" might eventually recover and regain the numbers of animals they had prior to chytridiomycosis or if these populations will remain small or even eventually disappear.

Although the exact origin of Bd *has not yet been determined, it has become clear that global trade in amphibians . . . is responsible for movement of* Bd *to locations where it was not previously present.*

Recent research has shown that a critical factor in determining if chytridiomycosis will cause extinction of an amphibian population is if the level of intensity of the infection with *Bd* crosses a certain threshold. What is very interesting about the "persistent" populations is that the remaining animals are still infected with *Bd*, but at a lower or less lethal intensity. Like the individual amphibian species that are resistant to chytridiomycosis (see above), understanding why persistent populations maintain low intensity infections with *Bd* is very important and could lead to methods to control the disease in wild populations.

Bd's Origin

Since its discovery, *Bd* has been found in wild and captive amphibian populations on every amphibian-inhabited continent. It is actively spreading in South, Central, and western North America, as well as the Caribbean, Australia and Europe. *Bd* is also found in Africa, Asia, and eastern North America, but does not seem to be spreading in these locations. *Bd* is conspicuously absent from Madagascar, Borneo and New Guinea.

Scientists have questioned if *Bd* is a fungus that has always infected amphibians all over the world and has just now begun to cause disease—because of changes to the environment or suppression of amphibian immune systems—or if *Bd* has only recently been introduced to new populations of amphibians and causes disease in naïve populations that have not developed a natural resistance to *Bd* infection. It is now well-documented for amphibian populations in Central America and the western United States that *Bd* was not present in the population until the beginning of declines due to chytridiomycosis. In other words it appears that *Bd* was newly introduced to these locations and then caused the population declines.

So if *Bd* has only recently been introduced to new locations, where did it come from? There is genetic and historical evidence that *Bd* has been present for a long time in Africa; Japan and eastern North America and all have been proposed as the possible site of origin. Although the exact origin of *Bd* has not yet been determined, it has become clear that global trade in amphibians for food, for use as laboratory animals, or for use as pets or display animals is responsible for movement of *Bd* to locations where it was not previously present. This has led to international regulations under the World Organization for Animal Health to require that amphibians be free of *Bd* infection before international shipment.

The Spread of *Bd*

Infection with *Bd* is transmitted by a form of the fungus called a "zoospore". The zoospore has a very distinctive appearance with a single flagellum that helps the spore swim through water or moist environments. Zoospores require moisture and cool temperatures and can persist in moist environments for several months, but do not tolerate conditions that are warm or dry for more than a few hours. Therefore, the most common and successful ways that *Bd* zoospores spread from place to place are in water, moist or wet materials (including soil or equipment) or on the skin of infected amphibians. In fact, the most common way that *Bd* infection spreads between amphibians is from direct contact of an infected animal with an uninfected animal (e.g. during territorial or breeding encounters). In captivity, it is possible to house amphibians infected with *Bd* in enclosures next to enclosures with amphibians that are not infected with *Bd* and not transmit the infection as long as animals, water and wet materials and tools are not shared between the enclosures. . . .

In the natural environment, it has been hypothesized that *Bd* can move on people's boots or equipment or on birds and invertebrates that fly between watersheds. Therefore, it is important that biologists and others take precautions to clean and disinfect their boots and equipment before moving from one location that has amphibians to another location in order to minimize the risk of spreading *Bd*. Because many amphibians that are infected with *Bd* are resistant to the disease chytridiomycosis (see above), they can appear to be outwardly healthy but are still capable of spreading *Bd* from one location to another. This is important because these animals may act as a reservoir for transmitting *Bd* infection to other amphibians as part of natural movements between different watersheds.

Amphibians can also move *Bd* to new locations as the result of trade in amphibians (see above) or potentially by the release of captive amphibians to the wild.

Identifying Chytridiomycosis

An amphibian that is sick with chytridiomycosis can have a wide variety of symptoms or "clinical signs". Some of the most common signs are reddened or otherwise discolored skin, excessive shedding of skin, abnormal postures such as a preference for keeping the skin of the belly away from the ground, unnatural behaviors such as a nocturnal species that suddenly becomes active during the day, or seizures. Many of these signs are said to be "non-specific" and many different amphibian diseases have signs that overlap with those of chytridiomycosis. In addition, other cases of chytridiomycosis will not show any of these signs and amphibians will simply be found dead. For these reasons it is not possible to diagnose chytridiomycosis with the naked eye and laboratory testing is required.

Diagnosing Chytridiomycosis

If animals are sick it is possible to diagnose chytridiomycosis by examining samples of the skin under a microscope and identifying the characteristic fungal organisms of *Bd*. These techniques require the assistance of an experienced biologist or veterinarian and are not good ways to detect amphibians that are carriers of *Bd*. Alternatively, non-invasive swabs of the skin can be obtained and analyzed by a technique called the polymerase chain reaction or "PCR" for short. PCR can detect very small amounts of *Bd* DNA in a sample and for this reason it is the test of choice for detecting animals that carry *Bd* infection and to survey wild and captive amphibian populations for the presence of *Bd*. . . .

PCR requires a molecular biology laboratory that uses rigorous controls for positive and negative samples and that has carefully validated the PCR test. A disadvantage of PCR is that it is not able to distinguish between amphibians that are sick with chytridiomycosis and amphibians that are carriers of *Bd*, because both types of animals will test "positive" by PCR.

Treating Chytridiomycosis

In captive amphibians, chytridiomycosis can be successfully treated with antifungal medications and by disinfection of contaminated enclosures. A variety of different antifungal medications have been described for the treatment of chytridiomycosis, however, one of the most common methods was developed at the Smithsonian National Zoo and uses a series of baths in the drug itraconazole. Itraconazole baths have been used successfully in rescue operations that capture wild amphibians from populations that are experiencing deaths to chytridiomycosis. Other potential treatment methods include the use of elevated body temperature and paradoxically, the antibiotic chloramphenicol. Treatment is not always 100% successful and not all amphibians tolerate treatment very well, therefore chytridiomycosis should always be treated with the advice of a veterinarian.

Unfortunately, there are no good methods for the treatment of wild animals in the natural environment. It is very difficult or impossible to get enough of the antifungal medications into the environment to be able to successfully rid infected frogs of *Bd*. In the future it may be possible to treat some amphibians in the wild in order to reduce the intensity of infection to a less lethal level with the hope that animals could survive with a mild *Bd* infection. Another promising area of research is looking at the possibility of introducing symbiotic bacteria that inhibit the growth of *Bd* into wild amphibian populations. So far, there is no evidence that a vaccine for chytridiomycosis could be effective for controlling the disease in wild populations.

Avoiding *Bd* and Chytridiomycosis

Amphibians are commonly kept in captivity as pets, laboratory animals, education animals and for species conservation efforts. In these situations, prevention and control of *Bd* infection and chytridiomycosis have become very important for maintaining healthy captive populations. Methods that are helpful in this regard include:

- Quarantine of new amphibians before they enter an established amphibian collection. New animals are kept separate from the established collection for a period of time (usually 60–90 days) to allow for observation for signs of disease and to perform laboratory testing for diseases such as *Bd*.
- Testing or treating animals for *Bd* infection during the quarantine period.
- Perform surveillance for *Bd* infection in your amphibian collection. This is done by regular necropsies (autopsies) of animals that die and by PCR testing of collection animals. Many amphibian collections have *Bd* infected frogs and don't know it.
- Develop "specific pathogen free" amphibian populations that are known to be free of *Bd* infection. If all captive raised amphibians can be certified as *Bd*-free it will simplify quarantine and amphibian shipment practices for everyone.
- Practice good hygiene and barrier management between animal rooms and displays. Use separate equipment and disposable gloves between enclosures and dispose of wastes and waste water responsibly.

If *Bd* is identified in your amphibian collection: DON'T PANIC. *Bd* infection is common in captive amphibians and there are effective treatment methods available (see above).

Use outbreaks of chytridiomycosis in collection as an opportunity to make your animals healthier by screening the collection for unsuspected carriers of *Bd* infection; treating infected animals and reviewing protocols for controlling the spread of infectious diseases in the collection. . . .

Habitat Loss

Habitat loss affects more amphibian species than any other threat by nearly a factor of 4. However, while habitat loss proceeds at a steady pace, *Bd* can often work quickly. The IUCN [International Union for Conservation of Nature] has called amphibian chytridiomycosis "the worst infectious disease ever recorded among vertebrates in terms of the number of species impacted, and its propensity to drive them to extinction." And because the Amphibian Ark [organization] focuses on species facing threats that cannot currently be mitigated in the wild, such as *Bd*, we necessarily focus largely on this disease and leave the mitigable threats, such as habitat loss, to our ASA [Amphibian Survival Alliance] partners specializing in those areas.

8

Global Warming May Force Polar Bears into Extinction

Michael McCarthy

Michael McCarthy is the environment editor for the Independent, *a national morning newspaper published in London, England.*

According to Canadian biologists, polar bears in the Hudson Bay area of Canada are likely to become extinct in the next three decades, or even sooner. As a result of global warming, the Arctic sea—ice where polar bears spend the winter hunting for seals—has begun to melt earlier and earlier, so that the animals do not have time to get the food they need to survive. Declining polar bear populations in Hudson Bay are perhaps the best indicator of the future of polar bear populations elsewhere, as the Hudson Bay is the second-most southern polar bear habitat, and thus expected to feel the effects of climate change early. Through studying the Hudson Bay polar bears, scientists have developed a mathmatical model to better predict the decline in the polar bear population.

Polar bears in the Hudson Bay area of Canada are likely to die out in the next three decades, possibly sooner, as global warming melts more Arctic ice and thus reduces their hunting opportunities, according to Canadian biologists.

Global Warming Threatens Polar Bears

The animals in western Hudson Bay, one of 19 discrete sub-populations of the species around the Arctic, are losing fat and body mass as their time on the floating sea ice gets shorter and shorter, according to the researchers from the University of Alberta.

The sea ice is where the bears hunt ringed and bearded seals, their main prey, and they have to build up enough fat in the winter, when the ice is at its greatest, to get through the summer, when the ice retreats from the shoreline and the bears can find no food.

The west Hudson Bay population has declined from 1,200 animals to 900.

But the ice has been melting earlier in the spring and forming later in the autumn, so that the bears are now spending on average three more weeks on land per year, without food, than they did three decades ago, the researchers say. As a consequence, their body weight in that time has dropped by 60lb, females have lost 10 per cent of their body length, and the west Hudson Bay population has declined from 1,200 animals to 900.

A Sign of the Future?

If the decline in the sea ice continues—as predictions of global warming suggest it will—it is feared that the bears could die out in 25 to 30 years, or perhaps in as few as 10, if there are a succession of years with very low sea ice cover. The Hudson Bay group of bears is the second-most southerly population and might be expected to feel the effects of climate change early. The Arctic sea ice as a whole reached its lowest-ever recorded extent in September, 2007. In the last two years it has recovered, but it is once again declining rapidly this year [2010].

The dependency of the bears on the ice has long been known, and the animals have become an iconic species in terms of being used to promote awareness of global warming.

But predictions of how long they may survive have until now been little more than educated guesses.

The significance of the new study is that it is based on a mathematical model which matches the weight and energy-storing capacity of the bears, which are known—the west Hudson Bay animals are the most closely observed of all polar bear populations—against the annual ice shrinkage and the time they have to spend on land without food.

The bears could die out in 25 to 30 years, or perhaps in as few as 10.

Carried out by Professors Andrew Derocher and Mark Lewis, with graduate student Peter Molnar, it has been published in the journal *Biological Conservation*, and Professor Derocher talks about it at length in the current issue of *Environment 360*, the online environmental journal of Yale University in the US.

"We understand very well things like how fat a bear has to be to produce a certain number of cubs, and we know a lot about how much energy these bears are burning during the period of time over the summer that they're forced ashore when the sea ice melts," Professor Derocher says. "And from there it's fairly easy to run various scenarios of sea ice change to look at when, basically, the bears' fat stores run out—and when that happens the bears, of course, subsequently die."

He adds: "There's been a gradual decline in [the bears'] body condition that dates to the 1980s and we can now correlate that very nicely with the loss of sea ice in this ecosystem. And one of the things we found was that the changes that could come in this population could happen very dramatically, and a lot of the change could come within a single year, if you just ended up with an earlier melt of sea ice."

9

Polar Bears: Today's Canaries in the Coal Mine?

Bjorn Lomborg

Bjorn Lomborg is an adjunct professor at the Copenhagen Business School in Denmark, the director of the Copenhagen Consensus Centre, and the author of The Skeptical Environmentalist.

Many news reports have said that polar bears are threatened with extinction by global warming. These reports are wildly exaggerated. In fact, research shows that polar bear populations are increasing worldwide. The main threat to polar bears is human hunting. The case of the polar bear illustrates a broader truth that the threat from global warming is overstated. Global warming is real, but it is not the worst problem facing the world. Responses to it should be kept in perspective.

Countless politicians proclaim that global warming has emerged as the preeminent issue of our era. The European Union calls it "one of the most threatening issues that we are facing today." Former prime minister Tony Blair of the United Kingdom sees it as "the single most important issue." German chancellor Angela Merkel has vowed to make climate change the top priority within both the G8 and the European Union in 2007, and Italy's Romano Prodi sees climate change as the real threat to global peace. Presidential contenders from John McCain to Hillary Clinton express real concern over the

Bjorn Lomborg, "Polar Bears: Today's Canaries in the Coal Mine?," *Cool It: The Skeptical Environmentalist's Guide to Global Warming*, New York, NY: Alfred A. Knopf, 2007, pp. 3–8.

issue. Several coalitions of states have set up regional climate-change initiatives, and in California Republican governor Arnold Schwarzenegger has helped push through legislation saying that global warming should be a top priority for the state. And of course, Al Gore has presented this message urgently in his lectures as well as in the book and Oscar-winning movie *An Inconvenient Truth*.

In March 2007, while I waited to give evidence to a congressional hearing on climate change, I watched Gore put his case to the politicians. It was obvious to me that Gore is sincerely worried about the world's future. And he's not alone in worrying. A raft of book titles warn that we've reached a *Boiling Point* and will experience a *Climate Crash*. One is even telling us we will be the *Last Generation* because "nature will take her revenge for climate change." Pundits aiming to surpass one another even suggest that we face medieval-style impoverishment and societal collapse in just forty years if we don't make massive and draconian changes to the way we live.

Likewise, the media pound us with increasingly dramatic stories of our ever worsening climate. In 2006, *Time* did a special report on global warming, with the cover spelling out the scare story with repetitive austerity; "Be worried. Be *very* worried." The magazine told us that the climate is crashing, affecting us both globally by playing havoc with the biosphere and individually through such health effects as heatstrokes, asthma, and infectious diseases. The heart-breaking image on the cover was of a lone polar bear on a melting ice floe, searching in vain for the next piece of ice to jump to. *Time* told us that due to global warming bears "are starting to turn up drowned" and that at some point they will become extinct.

Padding across the ice, polar bears are beautiful animals. To Greenland—part of my own nation, Denmark—they are a symbol of pride. The loss of this animal would be a tragedy. But the real story of the polar bear is instructive. In many ways, this tale encapsulates the broader problem with the climate-change concern: once you look closely at the supporting data, the narrative falls apart.

Al Gore shows a picture similar to *Time*'s and tells us "a new scientific study shows that, for the first time, polar bears have been drowning in significant numbers." The World Wildlife Fund actually warns that polar bears might stop reproducing by 2012 and thus become functionally extinct in less than a decade. In their pithy statement, "polar bears will be consigned to history, something that our grandchildren can only read about in books." *The Independent* tells us that temperature increases "mean polar bears are wiped out in their Arctic homeland. The only place they can be seen is in a zoo."

The heart-breaking image on the cover was of a lone polar bear on a melting ice floe, searching in vain for the next piece of ice to jump to.

Over the past few years, this story has cropped up many times, based first on a World Wildlife Fund report in 2002 and later on the Arctic Climate Impact Assessment from 2004. Both relied extensively on research published in 2001 by the Polar Bear Specialist Group of the World Conservation Union.

But what this group really told us was that of the twenty distinct subpopulations of polar bears, one or possibly two were declining in Baffin Bay; more than half were known to be stable; and two subpopulations were actually *increasing* around the Beaufort Sea. Moreover, it is reported that the global polar-bear population has *increased* dramatically over the past decades, from about five thousand members in the 1960s to twenty-five thousand today, through stricter hunting regulation. Contrary to what you might expect—and what was not pointed out in any of the recent stories—the two populations in decline come from areas where it has actually been getting colder over the past fifty years, whereas the two increasing populations reside in areas where it is getting warmer. Likewise, Al Gore's comment on drowning bears suggests an ongoing process getting ever worse. Actually, there was a single sighting of four dead bears the day after "an abrupt windstorm" in an area housing one of the increasing bear populations.

The global polar-bear population has increased *dramatically over the past decades.*

The best-studied polar-bear population lives on the western coast of Hudson Bay. That its population has declined 17 percent, from 1,200 in 1987 to under 950 in 2004, has gotten much press. Not mentioned, though, is that since 1981 the population had soared from just 500, thus eradicating any claim of a decline. Moreover, nowhere in the news coverage is it mentioned that 300 to 500 bears are shot each year, with 49 shot on average on the west coast of Hudson Bay. Even if we take the story of decline at face value, it means we have lost about 15 bears to global warming each year, whereas we have lost 49 each year to hunting.

In 2006, a polar-bear biologist from the Canadian government summed up the discrepancy between the data and the PR [public relations]: "It is just silly to predict the demise of polar bears in 25 years based on media-assisted hysteria." With Canada home to two-thirds of the world's polar bears, global warming will affect them, but "really, there is no need to panic. Of the 13 populations of polar bears in Canada, 11 are stable or increasing in number. They are not going extinct, or even appear to be affected at present."

The polar-bear story teaches us three things. First, we hear *vastly exaggerated and emotional claims* that are simply not supported by data. Yes, it is likely that disappearing ice will make it harder for polar bears to continue their traditional foraging patterns and that they will increasingly take up a lifestyle similar to that of brown bears, from which they evolved.

They may eventually decline, though dramatic declines seem unlikely. But over the past forty years, the population has increased dramatically and the populations are now stable. The ones going down are in areas that are getting *colder*. Yet we are told that global warming will make polar bears extinct, possibly within ten years, and that future kids will have to read about them in storybooks.

Second, polar bears are *not the only story*. While we hear only about the troubled species, it is also a fact that many species will do *better* with climate change. In general, the Arctic Climate Impact Assessment projects that the Arctic will experience *increasing* species richness and higher ecosystem productivity. It will have less polar desert and more forest. The assessment actually finds that higher temperatures mean more nesting birds and more butterflies. This doesn't make up for the polar bears, but we need to hear both parts of the story.

Nowhere in the news coverage is it mentioned that 300 to 500 bears are shot each year.

The third point is that *our worry makes us focus on the wrong solutions*. We are being told that the plight of the polar bear shows "the need for stricter curbs on greenhouse-gas emissions linked to global warming." Even if we accept the flawed idea of using the 1987 population of polar bears around Hudson Bay as a baseline, so that we lose 15 bears each year, what can we do? If we try helping them by cutting greenhouse gases, we can at the very best avoid 15 bears dying. We will later see that realistically we can do not even close to that much good—probably we can save about 0.06 bears per year. But 49 bears from the same population are getting shot each year, and this we can easily do something about. Thus, if we really want a stable population of polar bears, dealing first with the 49 shot ones might be both a smarter and a more viable strategy. Yet it is not the one we end up hearing about. In the debate over the climate, we often don't hear the proposals that will do the most good but only the ones that involve cutting greenhouse-gas emissions. This is fine if our goal is just to cut those gases, but presumably we want to improve human conditions and environmental quality. Sometimes greenhouse-gas cuts might be the best way to get this, but often they won't be. We must ask ourselves if it makes more sense to help 49 bears swiftly and easily or 0.06 bears slowly and expensively.

The argument . . . is simple.

1. *Global warming is real and man-made.* It will have a serious impact on humans and the environment toward the end of this century.

2. Statements about the *strong, ominous, and immediate consequences of global warming are often wildly exaggerated*, and this is unlikely to result in good policy.

3. *We need simpler, smarter, and more efficient solutions for global warming* rather than excessive if well-intentioned efforts. Large and very expensive CO_2 [carbon dioxide] cuts made now will have only a rather small and insignificant impact far into the future.

4. *Many other issues are much more important than global warming.* We need to get our perspective back. There are many more pressing problems in the world, such as hunger, poverty, and disease. By addressing them, we can help more people, at lower cost, with a much higher chance of success than by pursuing drastic climate policies at a cost of trillions of dollars.

These four points will rile a lot of people. We have become so accustomed to the standard story: climate change is not only real but will lead to unimaginable catastrophes, while doing something about it is not only cheap but morally right. We perhaps understandably expect that anyone questioning this line of reasoning must have evil intentions. Yet I think—with the best of intentions—it is necessary that we at least allow ourselves to examine our logic before we embark on the biggest public investment in history.

Many other issues are much more important than global warming.

We need to remind ourselves that our ultimate goal is not to reduce greenhouse gases or global warming per se but to improve the quality of life and the environment. We all want to leave the planet in decent shape for our kids. Radically reducing greenhouse-gas emissions is not necessarily the best way to achieve that.

I hope this book can help us to better understand global warming, be smarter about solutions to it, and also regain our perpsective on the most effective ways to make the world a better place, a desire we all share.

Notes

3 The European Union calls it "one of: (EU. 2001:208).

3 prime minister Tony Blair of the United Kingdom sees it as: (Blair, 2004b; Cowell, 2007).

3 German chancellor Angela Merkel has vowed to: (DW staff, 2006; Prodi, 2004).

3 Presidential contenders from John McCain to Hillary: (Buncombe, 2005).

3 Several coalitions of states have set up: (AP, 2006a; Pew Research Center, 2006).

3 And of course, Al Gore has presented: (Gore & Melcher Media, 2006).

4 A raft of books titles warn: (Gelbspan, 2004; Cox, 2005; Pearce, 2006).

4 Pundits--aiming to surpass one another even: (Bunting, 2006).

4 In 2006, *Time* did a special: (*Time*, 2006).

4 *Time* told us that due to global: "And with sea ice vanishing, polar bears—prodigious swimmers but not inexhaustible ones—are starting to turn up drowned. "There will be no polar ice by 2060,' says Larry Schweiger, president of the National Wildlife Federation. 'Somewhere along that path, the polar bear drops out' " (Kluger, 2006).

5 AI Gore shows a picture similar: (Gore & Melcher Media, 2006:146); see also (lredale, 2005).

5 The World Wildlife Fund actually warns that: (Eilperin, 2004).

5 In their pithy statement, "polar: (BBC Anon., 2005).

5 *The Independent* tells us: A 2°C rise is now unavoidable, and it will mean polar bears are wiped out in their Arctic homeland. The only place they can be seen is in a zoo" (McCarthy, 2006).

5 Over the past few years: (Berner et al., 2005; Hassol, 2004; Norris, Rosentrater, & Eid, 2002).

5 Both relied extensively on research published in: The World Conservation Union is also known as the IUCN; the Polar Bear Specialist Group website is http://pbsg.npolar.no/default.htm (IUCN Species Survival Commission, 2001).

5 But what this group really: The IUCN counts twenty groups, but most commentators mention about nineteen subpopulations (lUCN Species Survival Commission, 2001:22).

5 Moreover, it is reported that the: (Krauss, 2006).

5 Contrary to what you might expect&and: (Michaels, 2004). See springtime temperatures at (Przybylak, 2000:606).

6 Actually, there was a single sighting of: (Monnett, Gleason, & Rotterman, 2005).

6 That its population: (Harden, 2005; WWF, 2006).

6 Not mentioned, though: (Stirling, Lunn, & Iacozza, 1999:302), as confirnled by (Amstrup et al., 2006:slide 44; Rosing-Asvid, 2006).

6 Moreover, nowhere in the news coverage: (IUCN Species Survival Commission, 2001:22).

6 In 2006, a polar-bear: (Taylor, 2006).

6 Yes, it is likely that disappearing: The Arctic Climate Impact Assessment finds it likely that disappearing ice will make polar bears take up "a terrestrial summer lifestyle similar to that of brown bears, from which they evolved." It talks about the "threat" that polar bears would become hybridized with brown and grizzly bears (Berner et al., 2005:509).

7 In general, the Arctic Climate Impact: "While there will be some losses in many arctic areas, movement of species into the Arctic is likely to cause the overall number of species and their productivity to increase, thus overall biodiversity measured as species richness is likely to increase along with major changes at the ecosystem level" (Berner et al., 2005:997).

7 It will have less polar desert and: (Berner et al., 2005:998).

7 The assessment actually finds that: (Berner et al., 2005:256).

7 We are being told that the plight: (Eilperin, 2004).

7 We will later see that realistically: This is based on a simple model starting in 2000 with a population of one thousand, a reduction of 1.5 percent (fifteen bears first year), and the full Kyoto Protocol reducing global warming by about 7 percent in 2100 (Wigley, 1998).

10

Give a Bee a Home

Hazel Sillver

Hazel Sillver writes about gardening for the Times, *the* Guardian, *and other British publications.*

Many bees, including honey bees, are declining in population and may be facing extinction. This is bad not just for bees but also for humans, who rely on bees to pollinate many plants and important agricultural crops. The source of the bee decline is unclear, although it may be related to disease and parasites. Gardeners can help to preserve bee populations by putting hives in their gardens. They also can help by raising flowers and plants that bees use for food.

The beekeepers of Coventry are huddled around one of their hives at Ryton Gardens, in Warwickshire, headquarters of the charity Garden Organic. Dressed a little like astronauts, in protective white suits and hoods, they carefully lift one bee-coated frame after another to inspect them.

Creating Gardens for Bees

You might expect a bee-friendly organic garden to be a meadow dotted with dandelions and daisies, but Ryton is a series of well-kept beds with a herb garden, a rose garden and all the other trappings of formal horticulture. "A garden doesn't have to be a mess of wild flowers to seduce bees," says Peter Spencer, of the Warwickshire Beekeepers' Association. "It can be as neat and stylish as you like, but it must be planted with certain flowers." This is an excellent time to plant bee-friendly perennials, rich in easily accessible nectar and pollen, and get them established before winter sets in.

At Ryton, a designated bee garden showcases the plants bees like best. Sophisticates such as bergamot, gaura and the architectural globe thistle mingle with helenium, mallow and cobalt pools of cornflowers. The hexagonal garden has a glass viewing hive from which people can watch bees at work and try to spot the queen.

The main reason for these wonderful gardens is the plight of the honeybee, which, like several of our 250 bee species, is in danger of extinction.

Other gardens open to the public are following suit, putting bee gardens or apiaries in place to illustrate how to plant for nectar: RHS Hyde Hall, in Essex, has just installed hives; Cragside, in Northumberland, now plants with bees in mind; there's a new bee garden at the National Botanic Garden of Wales, Carmarthenshire; and in Hertfordshire, a bee garden and education centre, BuzzWorks, is being created.

The main reason for these wonderful gardens is the plight of the honeybee, which, like several of our 250 bee species, is in danger of extinction. "I lost 9 out of 12 hives last year," Spencer says. "Many beekeepers have a similar story." In fact, this year's harvest has been so poor—as many as one in three of Britain's 240,000 hives did not survive the winter and spring—that Rowse, the UK's leading honey producer, predicts supermarkets will have no English honey to sell by Christmas.

The reasons for the crisis are complex, but disease, changing climate—in particular this year's late spring and wet summer, which have kept the bees confined to their hives—and loss of habitat are all important factors. The varroa mite, which originated in Asia, has ravaged our wild bee populations and is now affecting domestic hives, where it weakens the bees, making it harder for them to survive winter.

Experts warn of the arrival of the deadly small hive beetle from America at any time, and there is also the threat of colony collapse disorder—bees mysteriously deserting hives—which has hit American apiaries hard. Avoiding imported honey and flowers may help, as these sometimes harbour disease, but the key to stemming the decline is research.

The Vital Bee

"The bee is vital," says Francis Ratnieks, Britain's only professor of apiculture. "It pollinates many plants—if it vanishes, so will a huge proportion of flowers and crops. A lot more must be done to help it. Research is the main way, but we have trouble with funding. Although the bee contributes £165m a year to the agricultural economy, the Government gives it only £200,000 of health-research funds in return." The British Beekeepers' Association is running a campaign to change this (visit www.britishbee.org.uk), but we can also help by becoming bee-minded gardeners. This means avoiding insecticides, which kill beneficial insects such as bees as well as pests, and cultivating their favourite flowers. If bees eat well, they are less prone to disease—and we'll benefit by having healthier and more abundant flowers and vegetables, all pollinated by them, as well as the pleasing sound of their hum.

So, what should we be planting? In general, bees like flowers of the rose family (which includes apple blossom and pyracantha), the daisy family and herbs with small flowers, such as mint, lavender and rosemary. Plant in drifts, rather than singly, as this makes it easier for them to find these nectar bars, and don't bother with double-flowered species, which have no nectar.

Some beekeepers believe bees favour blue plants—at Ryton, the blue of the lavender seems a magnet—but many yellow flowers also have pulling power. On our visit, the golden rod is covered in foragers, their heads buried in the flowers; only their back legs, dusted with collected pollen, are visible.

The bee is vital. . . . It pollinates many plants—if it vanishes, so will a huge proportion of flowers and crops.

We are being encouraged to keep bees, but many people are wary, concerned about swarms and stinging. "Take a course with a local beekeeping group before you buy hives," Spencer advises. "And put them in the right place—then it's safe." The ideal garden is big, with boundary hedges at least 6ft high: bees tend to fly at a constant height, so this will force them above head level and make them less of a nuisance in your neighbour's garden. They rarely sting unless provoked. To be on the safe side, though, a hive should be at least 16ft from a part of the garden you use regularly.

If keeping honeybees doesn't appeal, or you don't have the space, why not make your garden a haven for solitary bees and bumblebees? They are also important pollinators and, like honeybees, in decline. Garden centres sell cute-looking wooden bee homes, but experts turn up their noses at them, pointing out that bumblebees, which form smaller colonies than honeybees, are much more likely to nest underground or in a sheltered, undisturbed spot—underneath a garden shed, say—while solitary bees prefer to shelter in holes in wood.

The best way to help bees, Spencer says, is to provide a constant source of food: "They need food throughout the season, particularly in spring, to feed their young, and in autumn, to see them through hibernation". Certain flowers, including crocus, aubretia, thyme, heather and sedum, provide a richer source of nectar and pollen than others, so include some of these high-energy blooms in your plantings. Fruit trees offer invaluable blossoms in spring.

At the viewing hive in Ryton, Spencer points to a bee that is vigorously waggling her abdomen. "Look, she's telling them where she's just been—where the best food is in relation to the sun," he says. If you fill your beds and containers with nectar- and pollen-rich plants, your garden could also become a Michelin-starred feeding station for local bees.

The best way to help bees . . . is to provide a constant source of food.

The Bees' Needs

Bees love the orange flowers of California poppies (*Eschscholzia californica*), which light up a garden and bloom for months. They look great growing through gravel.

Angelica archangelica is a 6ft herb with white-green flowers that bees can't resist, and its architectural shape makes it perfect for the contemporary garden.

Copy Kailzie Gardens, near Peebles in Scotland, and create a pattern on your lawn by planting circular patches of daisies or white clover and mowing around them.

Crocuses provide essential food for the first bees to emerge in early spring. Plant them now in groups in a sunny spot, in pots (where they need gritty soil) or in the ground.

Include patches of campanulas and asters (which provide autumn food) or create miniature hedges of rosemary or lavender to edge your flowerbeds.

11

Bee Extinctions, and a Resulting Catastrophe, May Be Inevitable

Bill Schutt

Bill Schutt is an associate professor of biology at the C.W. Post Campus of Long Island University in New York and a research associate at the American Museum of Natural History.

Bees have been experiencing colony collapse disorder, or CCD, a phenomenon in which adult worker bees leave the hive, effectively destroying the colony. Researchers are unsure why this is happening, though possible explanations include a virus or fungal parasites, perhaps enabled by weakened bee immune systems caused by human breeding of bees. A drop in bee populations could have devastating consequences for humans, because bees pollinate many food crops and are a vital keystone species.

[A] mite causing *major* concern today is *Varroa destructor*, which preys on several types of bees, including honey bees (*Apis*) and bumble bees (*Bombus*). *Varroa* can be considered an invertebrate vampire because it feeds on hemolymph [a blood-like fluid in bees]. Since the bee's circulatory system doesn't function in gas transport, there is no oxygen-carrying hemoglobin, and as a result hemolymph lacks the red color of vertebrate blood. It is, however, a complex liquid containing a variety of hemocytes [cells involved in immune response in invertebrates], cells that carry out many of

Bill Schutt, *Dark Banquet: Blood and The Curious Lives of Blood-Feeding Creatures*, New York, NY: Crown Publishing, 2008, pp. 246–251. Illustrated by Patricia J. Wynne. Sub-headings added by Gale/Cengage.

the same functions as their leukocyte [vertebrate cells involved in immune response] counterparts—functions that include phagocytosis [the process by which cells engulf solid food particles] and a role in the immune response. There's even a hemocytic version of stem cells [cells that divide and differentiate].

Mites and Bees

Female mites enter bee nests (or hives) where they lay their eggs just before the brood chambers containing the developing bees are capped by the adult bees. The parasites feed on larval and pupal instars [that is, immature bees] as well as the emerging adult bees, which are also used for transportation. As with other arthropod parasites, as *Varroa destructor* feeds it can transmit viral and bacterial pathogens to its host.

Recently, the dramatic and nearly worldwide loss of honey bees has become a major concern not only within the beekeeping industry but also among farmers who raise the more than ninety commercial crops commonly pollinated by bees.[1] Colony collapse disorder (CCD, formerly known as fall dwindle disease) is characterized by the sudden departure of most of the adult worker bees from the hive, leaving behind the queen, a few young workers, and an abandoned brood of larvae and pupae. Although the cause of CCD is still under investigation, the list of potential suspects includes mites, bacteria, fungi, viruses, long-term exposure to substances like pesticides—especially neonicotinoids (chemicals that mimic the neurotoxic effects of the compound found in tobacco), and poor nutrition.[2] There is even a suggestion, albeit far-fetched, that cell phones are the causative agent.

1. These include apples, pears, blueberries, almonds, pumpkins, and squash.
2. The malnutrition hypothesis posits that bees forced to pollinate large monoculture farms are missing something in their diets in much the same way that a dog fed nothing but bread would experience physical harm from such a diet and eventually starve to death. On a related note, weather (e.g., drought) can also negatively affect pollen-producing plants, resulting in pollen that is deficient in the nutrients the bees require.

Cell Phones and Bees

In a pilot study published by the International Association of Agriculture Students, researchers at the University of Koblenz/Landau in Germany, placed cell phone handsets near four of eight beehives. They set out to measure hive-building behavior (by comparing before and after photographs of the hive chambers) as well as the tendency of the bees to return to their hives after they'd been captured, marked, and released some eight hundred meters away. Although the researchers reported that during the experiment "it became clear that both weight and area (of the hive) were developed better by non-exposed bees" statistical analysis "never showed a difference between exposed and non-exposed colonies." Oddly, in their Results section, the authors presented only *half* of their bee return data. They reported that in one exposed colony, only six of twenty-five test bees returned home within forty-five minutes, while in a second exposed colony, no bees returned. These incomplete findings triggered the publication of several articles (e.g., "Scientists Claim Radiation from Handsets Are to Blame for Mysterious 'Colony Collapse' of Bees," "Cell Phone Plague Obliterates Bee Colony," "Honey Bees Can't Call Home") purporting to inform readers of the dramatic new scientific developments. Typical was an editorial in the *Waco Tribune Herald* (April 16, 2007) in which the author stated that "a growing theory is that cell phones cause bees to become so disoriented that they cannot find their way back home."

There is even a suggestion, albeit far-fetched, that cell phones are the causative agent.

The original researchers were clearly not amused. According to Dr. Wolfgang Harst, the lead author, "This evolved as a case study for us in the new 'copy and paste' journalism." Harst slammed "the erroneous depiction of our study," from "faulty facts" about the study itself, to the claim that "handsets are to blame for 'colony collapse.'" He informed me that the follow-up study is set for publication in the journal *Environmental Systems Research* and that "although the findings are not so 'alarming' or 'breathtaking' as in 2005, the differences we found between the full irradiated and non-exposed bees were significant."

A number of researchers have published studies strongly suggesting that CCD is caused instead by a virus transmitted to bees (and/or activated) by *Varroa destructor*, the previously mentioned, hemolymph-sucking bee parasite.

Two closely related viruses have been implicated: Kashmir bee virus and Israeli acute paralysis virus.[3] These viruses are thought to be common infective agents within bee colonies (approximately eighteen bee viruses have been described) until stress or another problem (like *Varroa*) causes them to become epidemic and lethal.

Breeding and Viruses

"They've been selectively breeding different honey bee strains for years—for traits like mild temper, honey production, and resistance to mites," said Kim Grant, biologist and beekeeper. "It's certainly possible they've also bred in some things they *hadn't*, planned on—like susceptibility to some of these bee viruses or compromised immune systems."

Currently, scientists are trying to determine methods to stop the spread of CCD—many of which involve *Varroa*. These include the development of new miticides and the introduction of *Varroa*-resistant bees into European and American bee colonies. Clearly, though, beekeepers and farmers are taking CCD extremely seriously since the potential exists for a global nightmare should the world's bee populations disappear.

Scientist and *New York Times* best-selling author Dr. Charles Pellegrino, a polymath whose novel *Dust* took an

3. Researchers have identified both of these viruses in nearly all hives with CCD but not in control hives.

apocalyptic view of what would happen should the earth's insects go extinct, was less than optimistic about the ramifications of a honey bee extinction event.

They've been selectively breeding different honey bee strains for years. . . . It's entirely possible they've also bred in some things they hadn't *planned on.*

"So what do you think is causing this?" I asked him in the spring of 2007, as we sat on my favorite bench in Washington Square Park.

"The feeling from people I've talked to with the CDC [Centers for Disease Control and Prevention] is that weakened bee immune systems seem to be the issue here, with mite infestations more of a secondary symptom."

"What's compromising their immune systems—cell phones?"

There was a pause and Dr. Pellegrino frowned. "You're kidding me, right?"

I shrugged.

"Well, it's still a bit of a poser," he continued. "If it's a viral agent like they're saying—even something akin to 'bee AIDS'—then I'm not terribly worried. Viruses usually adapt very quickly to their hosts—and a bad parasite usually ends up dead, inside its dead host. A viral problem can be expected to quickly self-correct."

"You mean evolve into a nonlethal strain?"

"Right. But if it's a fungus weakening their immune systems, that could be much more problematic."

"Why's that?"

"Fungi adapt more slowly than viruses or bacteria. Plus they're resistant to all but the sorts of antimicrobial agents that would kill the bees as well as their parasites."

Bee Deaths

I figured it was time to bring out the big guns. "What would happen if all the bees went extinct because of CCD?"

Dr. Pellegrino gave a chuckle, but there was no humor in it. "It doesn't need to be a total extinction event. If bee deaths should reach 80 to 90 percent worldwide, I estimate that the earth's carrying capacity for human beings could be reduced, essentially overnight, from a maximum of twelve billion to about six billion—and we're at six point seven billion now."

"So you think the result would be . . .?"

"I think the result would be widespread famine and economic collapse, on a planet where the kamikaze mentality has already turned religious extremists into tigers sharpening plutonium claws."

Without the honey bee, Rome falls.

"Okay . . . but why the huge effect—the lack of bee-pollinated crops?"

"That's part of it, Bill. We'd be reduced to harvesting wind-pollinated crops like wheat and corn. But just as important, certain organisms are keystone species—basically nature's cascade points. Should they go suddenly extinct, or should their numbers be greatly reduced, then the entire system is affected. The honey bee is one of those keystones. Knock them down, near to extinction, and our civilization is gone in five years. Without the honey bee, Rome falls."

We sat there silently for a minute, watching the chess players dustered at tables near the park's southwestern entrance.

"Checkmate," I muttered.

Pellegrino gave another humorless laugh. "You got that right."

12

Human Beings Will Be Extinct Within One Hundred Years

The Daily Star

The Daily Star *is the leading English language newspaper in Lebanon and the Middle East.*

Professor Frank Fenner, an Australian scientist who helped cure smallpox, has predicted that humans will probably be extinct within 100 years. Fenner believes the human race will become extinct, along with many other species, due to climate change, overpopulation, and "unbridled consumption." Drawing parallels between today's situation and the fate of Easter Island in the first millenium AD, Fenner explains that overpopulation will result in increased competition for fewer resources. While Fenner is pessimistic about the future, other scientists remain optimistic.

Eminent Australian scientist Professor Frank Fenner, who helped to wipe out smallpox, predicts humans will probably be extinct within 100 years, because of overpopulation, environmental destruction and climate change.

Too Late

Fenner, who is emeritus professor of microbiology at the Australian National University (ANU) in Canberra, said homo sapiens will not be able to survive the population explosion and "unbridled consumption," and will become extinct, perhaps within a century, along with many other species. United Na-

"'The End's So Close!': End of Humanity in 100 Years!," *The Daily Star*, January 11, 2011. http://www.thedailystar.net/newDesign/index.php.

tions official figures from last year estimate the human population is 6.8 billion, and is predicted to pass seven billion next year [2012].

Fenner told *The Australian* he tries not to express his pessimism because people are trying to do something, but keep putting it off. He said he believes the situation is irreversible, and it is too late because the effects we have had on Earth since industrialization (a period now known to scientists unofficially as the Anthropocene) rivals any effects of ice ages or comet impacts.

Fenner said that climate change is only at its beginning, but is likely to be the cause of our extinction. "We'll undergo the same fate as the people on Easter Island," he said. More people means fewer resources, and Fenner predicts "there will be a lot more wars over food."

Climate change is only at its beginning, but is likely to be the cause of our extinction.

Easter Island

Easter Island is famous for its massive stone statues. Polynesian people settled there, in what was then a pristine tropical island, around the middle of the first millennium AD. The population grew slowly at first and then exploded. As the population grew the forests were wiped out and all the tree animals became extinct, both with devastating consequences. After about 1600 the civilization began to collapse, and had virtually disappeared by the mid-19th century. Evolutionary biologist Jared Diamond said the parallels between what happened on Easter Island and what is occurring today on the planet as a whole are "chillingly obvious."

While many scientists are also pessimistic, others are more optimistic. Among the latter is a colleague of Professor Fenner, retired professor Stephen Boyden, who said he still hopes awareness of the problems will rise and the required revolutionary changes will be made to achieve ecological sustainability. "While there's a glimmer of hope, it's worth working to solve the problem. We have the scientific knowledge to do it but we don't have the political will," Boyden said.

13

It Is Extremely Unlikely Human Beings Will Become Extinct

Tobin Lopes, et al.

Tobin Lopes and his coauthors are associated with the Global Energy Management Program at the University of Colorado, Denver.

Developing scenarios of human extinction can help to evaluate potential threats and plan for the future. The authors, therefore, developed three scenarios in which human life on earth ultimately ends. These scenarios included variables such as nuclear war, asteroid strikes, disease outbreaks, and deliberate terrorism. The authors were surprised to discover that it was very difficult to come up with scenarios that would mean the end of the entire human race. All scenarios ending in extinction required some element of human action, which suggested to them that the human race has the ability to prevent its own extinction.

We have developed three scenario frameworks that capture different paths to human extinction. While we have intended for each of these scenarios to be challenging, relevant and plausible, certainly, efforts moving forward would establish more facts and trends, and place a much more rigorous research agenda underneath the ideas presented here. Our scenario frameworks are as follows:

Tobin Lopes, Thomas J. Chermack, Deb Demers, Madhavi Kari, Bernadette Kasshanna, and Tiffani Payne, "Human Extinction Scenario Frameworks," *Futures*, vol. 41, 2009, pp. 732–737.

- Scenario framework 1: A journey of a thousand miles begins with a single step.
- Scenario framework 2: A road paved with good intentions.
- Scenario framework 3: Pouring salt in the wound.

We have developed three scenario frameworks that capture different paths to human extinction.

Each of these scenario frameworks features a combination of different events that have dramatic impacts on the human race and ultimately lead to extinction.

Scenario Framework 1: A Journey of a Thousand Miles Begins with a Single Step

In 2010, the pandemic began; it was generated from a highly virulent airborne virus that originated somewhere in the Pacific Rim. It quickly expanded to Asia through travel routes.

After only three years, the Asian population had suffered greatly. Fifteen percent of the population perished. Including non-Asians, the pandemic took 6% of the world's population. The continuing deaths led to a global economic crisis due to heavy restrictions on product distribution and travel. Businesses closed due to loss of merchandise. Schools began to close at alarming rates from fear of spreading the disease. Major public events virtually disappeared off the calendar. Professional sports leagues went on hiatus one after another. Theatres and museums closed.

Food shortages grew due to halted commerce; malnutrition and starvation increased. This led to greater illness and chaos as if a cloak of desperation had enveloped the human race. Taking advantage of the turmoil and heartache, terrorist groups began to escalate their actions.

In 2018, the United States entered into an armed conflict with Syria and Iran due to their increasing support of terrorist activities and pursued a build up of weapons. The elevated conflicts led to major disruptions in Middle East oil supplies. South American countries also stopped exporting oil to the U.S. leaving China as the world's largest consumer of oil. As the Chinese industrialists prospered, unhampered by environmental regulations, air pollution worsened. Worldwide the pandemic seemed to be under control; however, mortality rates remained high due to starvation, disease, and violence.

After four years of engagement in the Middle East, a crippled economy and lack of support forced the U.S. military to withdraw. Although weakened from the engagement, Iran did not halt its mission of destroying Israel. Within a few months, nuclear weapons were used by both sides. Iran emerged victorious however much of the Middle East was now wasteland.

With civilization in its last days, the final blow was struck. A meteor, approximately the size of the island of Oahu [Hawaii], struck the Atlantic Ocean.

Three more decades of increasing pollution and radiation resulted in global warming that caused severe disruptions in the ecosystem including drastic climate change. Melting icecaps led to coastal flooding. In 2050, the California coast became the Sierra-Nevada Mountains. As temperatures remained high, the insect populations grew at an accelerated rate. As cities and towns attempted to relocate further inland, insects followed. Mosquitoes, locusts, and other parasites drastically affected agriculture and spread more disease. Sickness, hunger and death touched every corner of the world. Mass chaos and violence became the norm. Governments were no longer able to provide basic services or maintain order. Either completely unscrupulous or religiously fanatical leaders emerged; neither felt restraint about using nuclear weapons.

With civilization in its last days, the final blow was struck. A meteor, approximately the size of the island of Oahu [Hawaii], struck the Atlantic Ocean. The resulting tidal waves, earthquakes, and changes in weather served to extinguish the remaining humans from the earth.

View from the end of the journey

My family is gone. I will join them soon. I am not sure why I decided to write this; it just did not seem right to not leave some sort of explanation. Although, I am certain no one is left to find it. I am not sure how long ago it was (time does not make sense anymore) when we convinced the children and their families to come with us to North Dakota. In the more populated areas disease, violence and warfare were quickly taking their toll. Major cities were bombed, and anyone in those areas died almost instantly. The real horror was the slow agonizing death of those who did not die from the initial blast but from radiation sickness. . . .

We decided that the only way to survive was to leave and try to escape to a sparsely populated area. We were not the only ones leaving the cities; practically anyone that had the means tried to go somewhere. We were pleased that relatively few seemed to select the same destination that we had. . . .

Initially the local communities were cooperative and bartering was common; but the more scarce resources became, the more everyone focused only on themselves and their families. Desperation was setting in and along with it came the fear and violence it breeds. One by one various family members died throughout our community. It did not take long to realize the symptoms of radiation poisoning. The horror we were most afraid of and tried to escape had caught up with us. We were all completely physically and mentally exhausted and overcome with grief and fear. The suffering reached an intolerable level, doing what I believed to be the most humane thing I could; I killed any wretched soul that remained alive. Now finally, my suffering also ends.

Scenario Framework 2: A Road Paved with Good Intentions

Early in the 22nd century a large pharmaceutical company, LoPhares, announced that after over six decades of work, a relatively inexpensive inoculation for cancer had been found. While it did not work for people who had already had cancer, it had been shown to be significantly effective for prevention of any type of cancer for those people who had not had cancer. LoPhares thought it was effective enough, in fact, to call it a vaccine. The only side effect was the slight blue skin discoloration (slightly larger than a U.S. quarter) where people received the shot. This discoloration typically appeared on the side of the neck just above the collar line.

The more prosperous countries and regions of the world were the first to use the vaccine that LoPhares called the Cancerous Hybrid Inoculation Prescription, or CHIP. By refining vaccination strategies almost two centuries old, it only took eight years for CHIP to reach the populations of the first and second world countries. Unstable regions in the Middle East and Central Africa that had volatile or suspicious leaderships made complete vaccinations almost impossible. With subsidization from the United Nations and other non-governmental agencies, vaccination in these areas continued at a much slower pace.

With only a small percentage of humanity vaccinated the cancer rate dropped almost to zero. 15 years after the first vaccinations, all but an estimated 400,000 people (besides children who were inoculated once they reached the age of two) in the unstable Central African regions had been vaccinated and cancer was relegated to historical texts. The United Nations had declared this point in history as The Last Year. Ten years later the last non-inoculated person in the world passed away. He was an 89-year-old retired computer programmer from Scranton, PA who lived most of his life in Arizona where he had had multiple instances of skin cancer. He died, tragically, when he was hit by a small child riding a bike. The impact dislodged a blockage in his heart that proceeded to his brain and killed him in a most alarmingly quick manner.

While the world was celebrating the success of CHIP, a sinister plot was unfolding. Answering the triumph of the inoculation, an ultra-patriotic environmental extremist (UPEE) began work to combat the byproduct of overpopulation. The UPEE utilized nanotechnology as his weapon of choice, and women's ovaries as the target. By designing a nano-attack intended to destroy the ovaries, the infection, subsequent to debilitating the host, became airborne, passing through the atmosphere to infect others. The method of dispersal was under the guise of an air purifier, slated for mass distribution to African, Asian, and Middle Eastern markets. A steady, if not alarming, decline in birth rates in the targeted markets exhilarated the UPEE, and he began to distribute the air purifiers to the Southern California and Atlanta, GA, markets under the guise of a smog deterrent. Because the air purifier was such a success, demand rose across the U.S., resulting in declining birth rates. Low birth rates across the world soon became a very real concern for humanity.

Seemingly, men were being affected by what was found to be a Hypercancer.

Sixty-six years after the death of the computer programmer, something unexpected happened. A Texas oil baron, son to two of the first vaccine recipients, passed away from a seizure-like event. An autopsy determined that his body was riddled with cancerous tumors. They had manifested themselves in record time, as he was the picture of health eight months earlier during a complete exam.

Seemingly, men were being affected by what was found to be a Hypercancer. Worldwide men started dying at alarming speeds. This new strand of cancer affected men at 100 times the rate that it affected women—it only took three years for Hypercancer to kill off 99.9% of the inoculated men in the world. The cause had yet to be discovered.

Journal entry: Year 69

In the beginning, it seemed as though everything was going well. I was so excited to be working with the lab that discovered The Cure. The company was doing extremely well. Hell, Pam, the kids, and I were doing extremely well too. We had a great house, great cars, great neighbors. The kids grew up with everything they wanted and needed, and even stuff they did not want nor need. . . .

I remember my retirement party. Pam, the kids, the grandkids, they were all there. It was great! A lot of my coworkers were there. Almost everybody I cared about was in the room. Little did I know it would be the last time I would see a great deal of the guys, including Steve, my youngest. Why did not we see this coming? Why did not we realize what we had done? We had done the lab work, followed research and methodology protocols. Yet it only took three years to undo all the good that we had accomplished. There were no men left. Only boys remained, and because of the decline in birth rates, there were very few. How could we know that The Cure was responsible for the Hypercancer? I guess, in our perch above humanity, we lost sight of the fact that there might be someone judging us. Ironic then that I, as one of the team leaders of The Cure, became a final witness of the human race. May God have mercy on my soul. May God forgive us.

Scenario Framework 3: Pouring Salt in the Wound

It was 1:35 AM aboard the Pacific Queen, a trans-oceanic cargo carrier. The deck was as busy as a beehive on this cold December night as the crew worked the ship's cranes and lifts.

Big black containers were carried and dumped into the ocean under the cover of darkness. The containers slammed into the ocean surface and then disappeared into the black depths. Everyone was working as fast as humanly possible, for it was not the best of situations to be caught dumping nuclear waste into the cold Atlantic Ocean in the middle of the thick night. Everyone aboard the ship was thinking the same thing: the sooner this task was finished, the sooner everyone could go home.

Six years later, people were being rushed to hospitals by the dozens. Old men, infants, women, everyone seemed to be victims of a strange phenomenon; all had died from internal hemorrhaging. Hospitals were overflowing with patients, and many were turned away and returned to their homes to face certain deaths. The doctors were helpless. They did not know what was causing these horrible mass deaths. Medical facilities were operating more as morgues rather than medical aid providers. Within hours of death, members of victim's families started dropping dead one after the other, a scene of insanity, of utter human tragedy.

No one seemed to understand what was going on. The disease seemed to target everyone at random without differentiating between age group, gender, or ethnicity. What made matters worse was that disease did not show any symptoms to speak of, it was more or less a sudden death situation for the victims. The government mobilized the National Guard and declared martial law in order to try to gain some control over the civil unrest that had begun. A curfew was imposed in an attempt to isolate the believed disease carriers and prevent further spread. Medical teams were called on the scene to examine the dead victims and central operation centers were established in state capitals and other large metropolitan areas.

Within three years, half the earth's population was taken out by this killer disease. The people under curfew died in their own houses one after the other not able to bury the dead, until the family home became the family graveyard. Soldiers, children, politicians, who wore biomedical suits were not spared.

After four years, scientists had their first breakthrough: the sudden death was being caused by bacteria that nested in the human blood stream and deteriorated the walls of the blood vessels, causing ruptures in the major arteries and rapid death due to extreme internal blood loss. As important a discovery as this was, it only brought more bad news. Apparently, the carrier of this strand of bacteria was regular table salt. The bacteria lie in the body for two to three years before multiplying and reaching a critical mass that resulted in a complete breakdown of arterial walls. Victims typically died within 13 h[ours]. The strand of bacteria was found to be a mutant form of a similar form of bacteria found in the common sea salt particles floating in the ocean and on dinner tables.

Within three years, half the earth's population was taken out by this killer disease.

The scientists disclosed the discovery of the source of this epidemic, but could provide no cure. They claimed that the salt purifying procedure was ineffective against these bacteria due to their endurance to extreme conditions. The official report was that anyone who had consumed salt within the last six months, from canned foods, restaurants or just plain sprinkles, had acquired the bacteria. If nothing could be done, they would perish. Unfortunately, it also appeared as if the bacteria were as highly adaptive as the Bird Flu. Once it entered the human body, it eventually found its way to the lungs where it mutated and became airborne. It started spreading through the air, not only through the consumption of salt. As a result, people could do nothing but await the inevitable; even the scientists working on the cure died one after the other. Six years after the initial infections, all the human population was dead leaving the Earth silent with nothing but the remnants of a civilization—just like an empty anthill. . . .

A final goodnight

It all started as a seemingly innocent disease that then grew into an epidemic that finally consumed the whole world. My husband, my two kids, and I spent many evenings watching in horror as the events unfolded. Death was everywhere. All we could see on the news were images of pandemonium . . . of chaos . . . of people dying . . . Nobody knew why. They called it The Disease; it was the end that we did not see coming. My husband was the first to succumb. He fell down on the floor convulsing in pain. Just like the other families, we tried taking him to the hospital. We were turned back by the National Guard.

My kids and I watched as the patriarch of the family withered away and died. We buried him in the backyard and promised him we would get over this terrifying situation. Two weeks later, my son Jeff fell ill and within two days of his death, my daughter Becky was gone. We were told that those who did not eat salt caught the virus through the air and it was going to be a while until we would catch up to our loved ones. With no electricity, no running water, no news, and the knowledge that everyone I knew was gone, I have grown very tired. I think it is time for me to go to sleep too. For that, I say one final goodnight.

Signposts

In exploring the implications of these scenario frameworks, we must first consider the signposts. Signposts are events that may be signaling the unfolding of a defined scenario. We see four key signposts that should draw serious consideration throughout the globe if they present themselves. Namely, (1) the launch of any nuclear weapon, (2) the outbreak of Avian flu or other disease with potential global impact, (3) questionable side effects of our tech/medical advancements, and (4) environmental responses. All but one have received prominent attention in the U.S. media.

Launch of nuclear weapons. This is an obvious concern for any individual, in any country on our planet. What is perhaps most compelling about this is not so much the event of any single strike, but rather, it is a combination of strikes in an "I'll get you before you get me" kind of mentality. With a number of states (e.g. Iran) advancing their development of nuclear technology the chance that an aggressive government or non-governmental organization gains access to such weaponry is very real.

Outbreak of disease. Current concerns about Avian flu outbreaks are real. Experts have suggested that it is not a matter of "if" this will happen, but "when" such an outbreak will occur. This thinking extends to other, as yet undiscovered, illnesses. Resources to deal with such an event certainly vary from country to country. To be sure, preparations for a global pandemic are under way, although the degree to which these threats are being taken seriously are debatable. Steps are also being taken to prepare for pandemics that are engineered for harm.

Side effects of technical/medical advancements. Much more subtle than the preceding two signposts, the side effects of our technological and medical breakthroughs should be seriously considered. Our scenarios have outlined the possibility that these side effects may only reveal themselves in time, which speaks to the delicate nature of advancements that we think will help, but in the long run, we are not quite sure. The signpost here will be abnormal and seemingly unexplainable medical problems, and a general increase in illness in post-industrial societies. Recent findings from cancer research indicate that some types of cancer could be caused by unknown environmental conditions. Perhaps such conditions include the additives in the food we eat, changes brought on by long-term exposure to magnetic or electronic fields, artificial cleansers, etc.

Environmental responses. Less within our control than the other key signposts we have outlined, environmental responses demand that we keep our planet in mind. Global warming, ozone depletion, and other natural responses to things like pollution and our technological advances applied to resolve pollution is just one example and something that has seen prolonged media attention. The key lesson here is simply trying to take the long view. Tools like systems theory can help us to understand more clearly the long-term implications of our actions.

We were very surprised about how difficult it was to come up with plausible scenarios in which the entire human race would become extinct.

Extinction Is Unlikely

These signposts also lead us to a few key conclusions that can be gleaned from our scenarios. First, the human race is unlikely to become extinct without a combination of difficult, severe and catastrophic events. We were very surprised about how difficult it was to come up with plausible scenarios in which the entire human race would become extinct. As we have stated, there was something reassuring about this fact.

We also speculated that in order for the human race to become extinct, it would have to be the result of some kind of intent, malicious or not, by some group of people, in some place of power on this planet. Naturally occurring phenomena can be devastating, but we found that we could not conceive of a naturally occurring phenomenon, or even a combination of them that would actually lead to our extinction. Thus, we learned that we, as a human race, have some influence over our ability to survive—we DO have a choice about how we solve our differences and explore solutions to highly complex, global problems. . . .

We are hopeful only that these frameworks might be used to provoke further dialogue on the topic of human extinction. While we found a tremendous amount of difficulty in creating plausible paths to extinction, the conversation itself was useful, engaging, and important. In our complex world with deeply volatile issues that could lead to an intent to create situations in which our race could become extinct, it is increasingly important to consider our responsibilities. We believe that exercises like this . . . are extremely influential and important in promoting the tools that we can use to better understand our world and how we can as individuals and groups, contribute more to it.

Organizations to Contact

The editors have compiled the following list of organizations concerned with the issues debated in this book. The descriptions are derived from materials provided by the organizations. All have publications or information available for interested readers. The list was compiled on the date of publication of the present volume; the information provided here may change. Be aware that many organizations take several weeks or longer to respond to inquiries, so allow as much time as possible.

Amphibian Ark (AArk)
c/o Conservation Breeding Specialist Group
12101 Johnny Cake Ridge Rd.
Apple Valley, MN 55124-8151
(952) 997-9800 • fax: (952) 997-9803
e-mail: KevinZ@amphibianark.org
website: www.amphibianark.org

The mission of Amphibian Ark is to ensure the global survival of amphibians, focusing on those that cannot currently be safeguarded in nature. The organization raises awareness of the global amphibian crisis, operates conservation training courses, develops partnerships among conservation groups, and raises funds for conservation. AArk's website includes news updates and extensive background on the amphibian crisis and efforts to combat it.

Conservation International
2011 Crystal Dr., Suite 500, Arlington, VA 22202
(703) 341-2400
website: www.conservation.org

Conservation International's goal is to promote biological research and work with national governments and businesses to protect biodiversity worldwide. On its website, the organization publishes fact sheets that explain its values, mission, strategies, and news and reports of its successes.

Earth Island Institute (EII)

300 Broadway, Suite 28, San Francisco, CA 94133-3312
(415) 788-3666 • fax: (415) 788-7324
website: www.earthisland.org

Founded in 1982 by veteran environmentalist David Brower, the Earth Island Institute develops and supports projects that counteract threats to the biological and cultural diversity that sustain the environment. Through education and activism, EII promotes the conservation, preservation, and restoration of the earth. In addition, the organization publishes the quarterly *Earth Island Journal.*

Environment Canada

351 St. Joseph Blvd., Place Vincent Massey, 8th Floor
Gatineau, Quebec K1A 0H3
Canada
(800) 668-6767 • fax: (819) 994-1412
e-mail: enviroinfo@ec.gc.ca
website: www.ec.gc.ca

Environment Canada is a department of the Canadian government. Its goal is the achievement of sustainable development in Canada through conservation and environmental protection. The department publishes reports, fact sheets, news, and speeches, many of which are available on its website.

Environmental Defense Fund

257 Park Ave. S, New York, NY 10010
(212) 505-2100 • fax: (212) 505-2375
website: www.environmentaldefense.org

Founded by scientists in 1967, the Environmental Defense Fund conducts original research and enlists outside experts to solve environmental problems. The advocacy group forms partnerships with corporations to promote environmentally friendly business practices. On its website, the organization publishes news, fact sheets, reports, and articles, including "The Importance of Wildlife and the Diversity of Life" and "Climate Change Could Be a Leading Cause of Biodiversity Loss."

Environmental Protection Agency (EPA)
Ariel Rios Building, 1200 Pennsylvania Ave. NW
Washington, DC 20460
(202) 272-0167
website: www.epa.gov

The Environmental Protection Agency is the federal agency in charge of protecting the environment and controlling pollution. The agency works toward these goals by enacting and enforcing regulations, identifying and fining polluters, assisting businesses and local environmental agencies, and cleaning up polluted sites. The EPA publishes speeches, testimony, periodic reports, and regional news on its website.

Friends of the Earth
1717 Massachusetts Ave. NW, Suite 600
Washington, DC 20036-2002
(877) 843-8687 • fax: (202) 783-0444
e-mail: foe@foe.org
website: www.foe.org

Friends of the Earth is a national advocacy organization dedicated to protecting the planet from environmental degradation; preserving biological, cultural, and ethnic diversity; and empowering citizens to have an influential voice in decisions affecting the quality of their environment. The organization publishes the quarterly *Friends of the Earth* newsmagazine, recent and archived issues of which are available on its website along with fact sheets, news, articles, and reports on environmental issues.

Greenpeace USA
702 H St. NW, Washington, DC 20001
(800) 326-0959 • fax: (202) 462-4507

e-mail: info@wdc.greenpeace.org
website: www.greenpeaceusa.org

Greenpeace USA is part of the worldwide Greenpeace organization, which opposes nuclear energy and the use of toxic chemicals and supports ocean and wildlife preservation. The organization uses controversial direct-action techniques and strives for media coverage of its actions in an effort to educate the public about environmental issues. It publishes the quarterly magazine *Greenpeace* and books, including *Coastline* and *The Greenpeace Book on Antarctica.* The organization's website features fact sheets; reports, such as *Food Security and Climate Change: The Answer Is Biodiversity*; and articles, such as "How Should We Protect Biodiversity and Our Climate?"

Natural Resources Defense Council (NRDC)
40 W 20th St., New York, NY 10011
(212) 727-2700 • fax: (212) 727-1773
e-mail: nrdcinfo@nrdc.org
website: www.nrdc.org

The Natural Resources Defense Council is a nonprofit organization that uses both law and science to protect the planet's wildlife and wild places and to ensure a safe and healthy environment for all living things. NRDC publishes the quarterly magazine *OnEarth* and the bimonthly bulletin *Nature's Voice.* On its website, NRDC provides links to specific environmental topics and news, articles, and reports, including *The Cost of Climate Change* and *Keeping Oceans Wild.*

Property and Environment Research Center (PERC)
2048 Analysis Dr., Suite A, Bozeman, MT 59718
(406) 587-9591
e-mail: perc@perc.org
website: www.perc.org

The Property and Environment Research Center is a nonprofit research and educational organization that seeks market-oriented solutions to environmental problems. PERC holds a variety of conferences about environmental issues and economics. It publishes the quarterly newsletter *PERC Reports*, commentaries, research studies, and policy papers, many of which are available on its website, including "Conserving Biodiversity through Markets: A Better Approach," "Meet the Enviropreneurs of 2008," and "Is There a Biodiversity Jackpot?"

Bibliography

Books

James P. Collins and Martha L. Crump	*Extinction in Our Times: Global Amphibian Decline*. New York: Oxford University Press, 2009.
Rebecca Costa	*The Watchman's Rattle: Thinking Our Way Out of Extinction*. Philadelphia, PA: Vanguard Press, 2010.
Richard Ellis	*On Thin Ice: The Changing World of the Polar Bear*. New York: Knopf, 2009.
Caroline Fraser	*Rewilding the World: Dispatches from the Conservation Revolution*. New York: Metropolitan Books, 2009.
Terry Glavin	*The Sixth Extinction: Journeys Among the Lost and Left Behind*. New York: Thomas Dunne Books, 2006.
Tony Hallam	*Catastrophes and Lesser Calamities: The Causes of Mass Extinctions*. New York: Oxford University Press, 2004.
Bjorn Lomborg	*Smart Solutions to Climate Change: Comparing Costs and Benefits*. New York: Cambridge University Press, 2010.
Alanna Mitchell	*Seasick: Ocean Change and the Extinction of Life on Earth*. Chicago, IL: University of Chicago Press, 2009.

Martin Rees *Our Final Hour: A Scientist's Warning*. Cambridge, MA: Perseus Books, 2003.

Michael Schacker *A Spring Without Bees: How Colony Collapse Disorder Has Endangered Our Food Supply*. Guilford, CT: Pequot Press, 2008.

Larry J. Schweiger *Last Chance: Preserving Life on Earth*. Golden, CO: Fulcrum, 2009.

Craig B. Stanford *The Last Tortoise: A Tale of Extinction in Our Lifetime*. Cambridge, MA: Belknap Press, 2010.

William Stolzenburg *Where the Wild Things Were: Life, Death, and Ecological Wreckage in a Land of Vanishing Predators*. New York: Bloomsbury USA, 2008.

Peter D. Ward *Under a Green Sky: Global Warming, the Mass Extinctions of the Past, and What They Can Tell Us About Our Future*. New York: Collins, 2008.

Edward O. Wilson *The Future of Life*. New York: Vintage Books, 2002.

Periodical and Internet Sources

ABC News "To Bee Or Not To Bee?: Colonies Face Extinction," April 24, 2009. www.abc.net.au.

David Biello — "Mass Extinctions Tied to Past Climate Changes," *Scientific American*, October 24, 2007. www.scientificamerican.com.

Charles Q. Choi — "Greatest Mysteries: What Causes Mass Extinctions?" LiveScience, August 8, 2007. www.livescience.com.

Charles Q. Choi — "Can the 'Amphibian Ark' Save Frogs from Pollution/Extinction?" *Scientific American*, June 19, 2008. www.scientificamerican.com.

DiscoveryNews — "20 Percent of Plant Species Face Extinction," September 29, 2010. http://news.discovery.com.

Niall Firth — "Human Race 'Will Be Extinct Within 100 Years,' Claims Leading Scientist," *Daily Mail*, June 19, 2010. www.dailymail.co.uk.

Richard Girling — "Plight of the Humble Bee," *Times*, February 1, 2009. www.timesonline.co.uk.

Independent — "World's Sixth Mass Extinction May Be Underway," March 7, 2011. www.independent.co.uk.

Elizabeth Kolbert — "The Sixth Extinction?" *New Yorker*, May 25, 2009.

Stephen Leahy — "Mass Extinction Not Inevitable," *Wired*, March 20, 2004. www.wired.com.

Robin McKie	"Britain's Birds Facing Extinction as Climate Change Leaves Them with Nowhere to Go," *Guardian*, March 1, 2009. www.guardian.co.uk.
John Muchangi	"Kenya: Indigenous Plants Face Extinction," *Nairobi Star*, April 22, 2011. http://allafrica.com.
Jeff Poor	"Climate Expert: Kyoto Would Save Only One Polar Bear a Year," Business & Media Institute—Media Research Center, June 25, 2008. www.mrc.org.
Andrew Revkin	"Vanishing Frogs, Climate, and the Front Page," *New York Times*, March 24, 2008. http://dotearth.blogs.nytimes.com.
John Roach	"Most Polar Bears Gone by 2050, Studies Say," *National Geographic*, September 10, 2007. http://news.nationalgeographic.com.
Ian Sample	"Human Activity Is Driving Earth's 'Sixth Great Extinction Event,'" *Guardian*, July 28, 2009. www.guardian.co.uk.
Scientific American	"Freeze Out: Can Polar Bears Survive a Melting Arctic?" April 29, 2011. www.scientificamerican.com.
Telegraph	"Human Expansion Leading to 'Extinction Crisis,' UN Warns," January 11, 2010. www.telegraph.co.uk.

United Press International	"Fungus Pushing Amphibians to Extinction," May 3, 2011. www.upi.com.
Jennifer Viegas	"Human Extinction: How Could It Happen?" *DiscoveryNews*, November 1, 2009. http://news.discovery.com.

Index

A

B

C

D

E

F

G

H

I

J

K

L

M

N

O

P

Q

R

S

T

U

V

W

Y

Z